VILLAS IN
SOUTHEAST ASIA

东南亚风情别墅

深圳市艺力文化发展有限公司 编

华南理工大学出版社

·广州·

图书在版编目（CIP）数据

东南亚风情别墅 / 深圳市艺力文化发展有限公司编 . — 广州：华南理工大学出版社，2015.9

ISBN 978-7-5623-4678-4

Ⅰ．①东⋯　Ⅱ．①深⋯　Ⅲ．①别墅－建筑设计－图集　Ⅳ.
① TU241.1-64

中国版本图书馆CIP 数据核字（2015）第 133783 号

东南亚风情别墅
VILLAS IN SOUTHEAST ASIA
深圳市艺力文化发展有限公司　编

出 版 人：	韩中伟
出版发行：	华南理工大学出版社
	（广州五山华南理工大学 17 号楼，邮编 510640）
	http://www.scutpress.com.cn　E-mail: scutc13@scut.edu.cn
	营销部电话：020-87113487　87111048（传真）
责任编辑：	李 欣　王 岩
印 刷 者：	深圳市福威智印刷有限公司
开　 本：	615 mm × 1020 mm　1/16　　**印张**：20
成品尺寸：	248 mm × 290 mm
版　 次：	2015 年 9 月第 1 版　2015 年 9 月第 1 次印刷
定　 价：	360.00 元

版权所有　盗版必究　　印装差错　负责调换

PREFACE
序言

This is one of the first books of its kind; an aim towards a comprehensive detailing of the styles, structures and characteristics that make up Southeast Asia's luxury villas. The region has an exponentially expanding market for high-end holiday rental properties, and although there are far too many to list in one place, this book looks into the features that unite and separate these stunning estates.

The major talking points in looking at villas in this region are nature and luxury, and how architecture and décor reflect these two ideals. The natural and luxurious are regularly fused into design, often with incredible attention to detail; furnishings are bespoke and hand-picked for specific locations, whilst gardens and terraces are perfectly sculptured to offer the best surrounding views for every angle.

Southeast Asia is a part of the world that is renowned for its stunning natural beauty; from its beaches to its rock faces, to its greenery, and everything in between, natural beauty can be found everywhere. Due to this, these properties often try to encompass their surroundings, making the villa part of the environment, and using natural landscaping and materials where possible, in order to allow this natural beauty to permeate their design.

Luxury is also one of the main considerations taken into account when designing and furnishing these properties, ensuring that these villas are fit for the weary traveller; you will find every comfort imaginable. From infinity pools with built-in bars, to gyms, saunas, spas, to outdoor terraces with surround sound music systems, these villas create luxuriant high-end holiday experiences.

Not to mention the in-villa services; it is difficult to portray with images, but the villas in Asia are renowned for their hospitality, providing guests with outstanding quality in an array of different services; from housekeeping to creating sumptuous meals, to making cocktails and providing in-house spa treatments; these villas have it all.

I trust these properties will inspire your next villa holiday.

Marc Ribail
Founder and managing partner, Unique Retreats

本书作为首批介绍东南亚别墅的书籍之一，旨在详尽地分析东南亚风情豪华别墅的风格、结构及特色。虽然现有的高档别墅数不胜数，但东南亚对高端度假房屋租赁仍有很大需求，市场前景广阔。因此，本书深入探究了这些绝佳别墅的共性与个性。

此次探究的切入点是东南亚风情别墅的自然性和奢华性，并对建筑和装饰是如何体现这两点作了分析。对于东南亚的别墅而言，设计常常会融入自然与奢华两大元素，对细节的关注也总是令人不可思议，精心定制的室内陈设、悉心挑选的细节点缀、完美打造的花园和露台，每个视角都可呈现最佳的风景。

东南亚以自然风光之美而闻名于世，从美丽的海滩到险峻的岩石，再到葱郁的绿地，乃至这里的一切，无一不能体现自然之美。因此，这里的建筑试图与周围环境相融合，让别墅成为环境的一部分，尽可能地利用自然景观和天然材料，以求自然之美渗透到设计中。

在房屋的设计和装饰过程中，奢华也是考虑的重点之一，以确保别墅能满足疲惫旅客之需。在这里，一切能想象到的舒适场景，你都能找到。从内设酒吧的无边水池，到健身房、桑拿房、水疗室，再到覆盖着音响系统的户外露台，此类别墅能带来最奢华的度假体验。

别墅内更有周到至极的服务，这远非几张图片能呈现的。亚洲的别墅以其盛情的款待而著称，它为顾客提供一系列品质卓越的服务，从客房服务到豪华餐宴，从美味鸡尾酒到自带水疗护理，各种顶级餐食和服务应有尽有。

我相信书中的别墅定会为你的下一次别墅度假之旅带来启发。

Marc Ribail
Unique Retreats 创始人兼执行合伙人

CONTENTS
目录

001 /	SUMMARY OF SOUTHEAST ASIAN STYLE 东南亚风格概述		180 /	VILLA RAK TAWAN 瑞克塔湾别墅
008 /	VILLA MALAATHINA 玛拉提娜别墅		190 /	BULGARI VILLA 宝格丽别墅
022 /	PRESIDENTIAL VILLA 总统别墅		196 /	VILLA AMANZI 阿曼兹别墅
032 /	ABSOLUTE WATERFRONT VILLA IN CAPE YAMU 雅慕角海滨别墅		204 /	VILLA YIN "阴"别墅
042 /	BREATHTAKING SEA VIEW VILLA 令人惊叹的海岸别墅		214 /	VILLA NAMASTE 纳玛斯蒂别墅
052 /	VILLA SAMAKEE 萨马基别墅		222 /	VILLA BENYASIRI 本雅思瑞别墅
062 /	VILLA KYA 基亚别墅		232 /	VILLA FAH SAI 法萨伊别墅
074 /	VILLA BAAN WANORA 班瓦诺拉别墅		240 /	VILLA HALE MALIA 黑尔·玛丽亚别墅
086 /	VILLA BELLE 贝拉别墅		250 /	BAN MEKKALA 班麦卡拉别墅
098 /	VILLA KALYANA 卡雅拿别墅		260 /	VILLA YANG SOM 杨孙别墅
110 /	VILLA LEELAVADEE 里拉瓦迪别墅		272 /	VILLA VIMAN 维曼别墅
118 /	ICONIC ESTATE 雅致房产		280 /	VILLA ROM TRAI 罗泰别墅
130 /	VILLA SHAMBALA 香巴拉别墅		290 /	MALIMBU CLIFF VILLA 玛里姆布悬崖别墅
146 /	ESHARA VILLAS 伊莎拉别墅		298 /	VILLA LUWIH 露维别墅
158 /	VILLA LIBERTY 自由别墅		304 /	BAAN HINTA 巴安欣塔别墅
168 /	VILLA PADMA 帕德玛别墅		312 /	CONTRIBUTORS 设计师名录

SUMMARY OF SOUTHEAST ASIAN STYLE

东南亚风格概述

Southeast Asian Style is the design that combines the ethnic characteristics in the Southeast Asian islands and refined cultural tastes. It inherits the qualities of nature, health and recreation, and shows its respect to nature and advocates handicrafts from the whole space design to the subtle decorating details. From splendid traditional palatial architecture, colorful interior decoration, to the distinctive furniture design, all give out the unique and mysterious charm.

东南亚风格是一种结合了东南亚岛屿民族特色及精致文化品位的设计，继承了自然、健康和休闲的特质，大到空间打造，小到细节装饰，都体现了对自然的尊重和对手工艺制作的崇尚。从金碧辉煌的宫殿式传统建筑、缤纷绚丽的室内装饰，到独具一格的家具设计，无不透露出独特而神秘的风情。

The greatest feature of Southeast Asian architectures is the attention to sunshade, ventilation and daylighting etc. What's more, the selection of materials is of greater representativeness, for example, the use of yellow wood, bluestone otter board, cobblestone and granite and so on. Concerning the color solution, they mainly use religious dark colors, such as brown, pearl black, golden, bright pink and yellow etc., which appear calm and atmospheric. Those architectures under the influences of Western design styles are commonly in light colors, such as pearl color and creamy white, creating the soft feeling.

东南亚建筑的最大特色是对遮阳、通风、采光等条件的关注；对建筑材料的使用也颇具代表性，如黄木、青石网板、鹅卵石和花岗岩等；在用色上则主要以宗教色彩浓郁的深色系为主，如深棕色、黑珍珠色、金色和鲜艳的桃红色和黄色等，令人感觉沉稳大气。受到西式设计风格影响的那些建筑，则以浅色系比较常见，如珍珠色和奶白色等，给人以轻柔的感觉。

As for home furnishing, the obbligato tools contributing to the Southeast Asian style include cool cane chairs, Thai silk bolsters, exquisite woodcarvings, vivid Buddha's hands, enchanting gauze curtains and artful candlesticks etc. Even they are randomly placed, they still add the mystique to the space.

而在家居方面，清凉的藤椅、泰丝抱枕、精致的木雕、造型逼真的佛手、妩媚的纱幔和设计巧妙的烛台等都是成就东南亚风情最不可缺少的元素，即使随意摆设，也能平添几分神秘气质。

What is a must to mention is that garden design is an indispensable part of the architecture here, due to the Southeast Asian hot climate which is suitable for outdoor activities. Rebuilding the most natural feeling is known as the greatest feature of Southeast Asian gardens, through the full use of plants, tables, chairs and stones and other local materials, thus to achieve a simple and comfortable retreat ambience.

不可不提的是，由于东南亚长年气候炎热，适宜户外活动，故园林设计必然成为建筑里面不可缺少的部分。据了解，东南亚园林最大的特点是还原最自然的风情，充分运用当地材料，就如植物、桌椅、石材等都取材于当地，强调简朴、舒适的度假风情。

CHARACTERISTICS OF SOUTHEAST ASIAN ARCHITECTURES

东南亚建筑风格特点

▶ UNIQUE ISLAND LANDSCAPE
独特的岛屿景观

There are so many unique islands in the Southeast Asian region which form unique island landscapes, contributing to the unique Southeast Asian Style.

东南亚地区拥有众多独特的群岛，这些岛屿形成了独特的风景，成就了独特的东南亚风格。

▶ RICH RELIGIOUS COLORS
浓郁的宗教色彩

People in Southeast Asia believe in different religious, which also influences their architectures in the region, thus the architectures are full of religious features. Such architectures include not only churches, but also residences. Therefore, their home environment and garden design reveal unique religious characteristics.

东南亚地区的人信奉不同的宗教，这样的宗教信仰也在该地区的建筑中体现了出来，使其拥有浓郁的宗教特色。这样的建筑不仅包括教堂，还包括住宅，在家居环境设计以及园景设计中同样体现了特殊的宗教特色。

KEY POINTS OF SOUTHEAST ASIAN HOME DESIGN
东南亚家居设计要点

▶ ORIGINALITY IN SELECTION OF NATURAL MATERIALS
取材自然：别开生面

Southeast Asian furniture is mostly created with local materials. They are eco-friendly and stylish, with a timber hue and a natural and plain visual sense. For example, rattan, seaweed, coconut husk, shell, bark and sandstone alike are all usable in making furniture, lamps, lanterns and decors, full of strong natural flavor.

　　东南亚家具时尚环保，大多就地取材，家具以原木色调为主，在视觉上给人自然、质朴的感觉。比如，藤、海草、椰子壳、贝壳、树皮、砂岩石等之类的都可以拿来制作家具、灯具和饰品，散发着浓烈的自然气息。

▶ ECO DECORATIONS WITH UNDERSTATED ZEN SENSE
生态饰品：拙朴禅意

Decors are made by hand with natural rattan and teak, plain but of profound Zen sense. These eco decors are quite eye-opening. Their textures and colors go with natural beauty which cannot be achieved by man-made products.

　　饰品大多以纯天然的藤竹柚木为材质，纯手工制作而成，带着几分拙朴，却仿佛隐藏着无数的禅机。这些生态饰品让人大开眼界，其色泽纹理有着人工无法达到的自然美感。

▶ EMBELLISHING WITH WARM COLOR FABRICS
布艺饰品：暖色点缀

Various colorful fabrics are the best match for the Southeast Asian furniture, which avoids the monotonousness of furniture but adds vigorousness. As for the color selection of fabrics, the symbolic shining color of Southeast Asian style is a range of dark colors that can change under lights, calm but also luxury.

各种各样色彩艳丽的布艺装饰是东南亚家具的最佳搭档，用布艺装饰适当点缀能避免家具的单调气息而令气氛活跃。在布艺色调的选用上，东南亚风情标志性的炫色系列多为深色系，且在光线下会变色，沉稳中透着一点贵气。

▶ COLORFUL LUXURY
色彩搭配：斑斓高贵

Decorating with dramatic and bright colors to eliminate the visual tediousness, Southeast Asian home design emphasizes that colors should return to nature since bright colors are the colors of nature.

装饰中运用夸张艳丽的色彩打破视觉的沉闷，斑斓的色彩其实就是大自然的色彩，色彩回归自然也是东南亚家居的特色。

CHARACTERISTICS OF SOUTHEAST ASIAN GARDENS
东南亚园林风格的特点

▶ HARD LANDSCAPE CHARACTERISTICS
硬质景观特点

• GROUND PAVING
地面

The paving of grounds in Southeast Asian landscape is delicate but never over done. There are no special requirements in material selection. But if clients require highlighting the paving effect in the courtyard, there will be some striking patterns.

东南亚景观的道路铺装精致但不过度，材料选择上没有特别要求。如果需要强调庭园铺装效果，可以铺设一些醒目的图案。

• PAVILION, TERRACE AND BRIDGE
亭、榭、桥

It is not a real tropical garden without pavilion, bridge or terrace. With exotic decorating structures added, a delicate garden can be more elegant.

没有凉亭、桥和榭台的景观就不算是真正的热带园林，通过添加有异国情调的装饰性构筑物，可以使一座精美的庭园更加美轮美奂。

GARDEN ORNAMENT

园林小品

The common garden ornament of Southeast Asian style includes sculpture, hummock, landscape ornament, path, stepping stones, paving and garden etc. These elements are usually the finishing touches of Southeast Asian Style garden, with strong decorating effects.

东南亚风格中常见的园林小品包括：雕塑、小丘、植物造景、路径、踏脚石、铺砖和园景等，这些元素在东南亚风格的景观中往往是点睛之笔，具有很强的装饰作用。

WATER LANDSCAPE CHARACTERISTICS

水景景观特点

Irregular water scenery is often used in the Southeast Asian palaces. These gardens decorated with water landscapes are deeply influenced by Eastern, Western aesthetical traditions and cosmology. Southeast Asian water landscapes respect Nature, with rich layers and levels. The transition from water surface to the ground is natural, and the aquatic plants, floating plants and emergent aquatic plants match up with the pebbles, sands, sculptures.

东南亚各地的宫廷建筑中，大体量地采用不规则的水体，这些水景庭园深受东西方美学传统和宇宙观的影响。东南亚水景景观崇尚自然，立面层次丰富，从水面到地面的过渡自然，水生植物、浮生植物以及挺水植物与卵石、素沙、雕塑相配。

SOFT LANDSCAPE CHARACTERISTICS

软质景观特点

The Southeast Asian style gardens often use greenery to highlight the tropical aroma. Greenery is the "necessary" keys that designers usually use to create Southeast Asian style garden, such as the tropical tall palms and vines which bring the best effect. The common tropical plants are Phoenix dactylifera, palm, coconut tree, scindapsus, sago cycas, rubber tree, jack fruit and bougainvillea, etc. All are of strong tropical flavor.

在东南亚风格景观园林中，绿色植物也是突显热带风情关键的一笔，尤其以热带大型的棕榈树及攀藤植物效果最佳，目前最常见的热带植物有海枣、华棕、椰子树、绿萝、铁树、橡胶树、菠萝蜜、勒杜鹃，等等，这些植物极富热带风情，是设计师常用来营造东南亚风格景观园林的"必备"品。

VILLA MALAATHINA
玛拉提娜别墅

Contributor:
Marketing Villas Ltd.

Location:
Umalas village, Seminyak, Bali

Area:
5,000 m² (Land)

Prepare to live like royalty at this extraordinary seven-bedroom villa. Located in the charming, traditional village of Umalas, near Seminyak, and surrounded by rice paddies, this villa is sumptuous even by the incredibly high standards of contemporary Balinese accommodation.

Set on an enormous piece of land, this splendid modern villa ticks all the boxes for the ultimate Bali escape: seven luxuriously appointed bedrooms; a 25-metre swimming pool surrounded by sun lounges on which to while away a lazy day; an air-conditioned library where you can take morning coffee or sip a cooling cocktail; a state-of-the-art gym; a yoga studio overlooking rice paddies, and even a Teppanyaki pavilion where your chef will create meals to rival the finest Japanese restaurants.

As well as the in-house chef, Villa Malaathina's large team of skilled and professional staff includes an experienced villa manager, housekeepers, service and security staff, whose aim is to ensure you want for nothing. And should you wish to celebrate a special birthday, anniversary or even a wedding during your stay, the villa's spacious grounds will provide the perfect setting for a truly memorable occasion.

Villa Malaathina provides you with the best that beautiful Bali has to offer, and it's hard to imagine why anyone would want to leave this luxurious holiday abode, but if it's chill-out beach bars, fine-dining restaurants and hip boutiques that you're after, then buzzing Seminyak has it all, and is a mere fifteen-minute drive away.

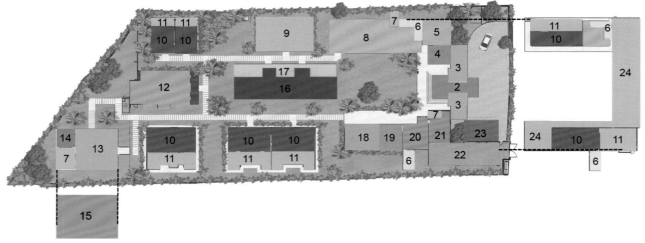

KEY PLAN

1. Entrance
2. Villa Entrance
3. Pond
4. Study
5. Media Room
6. Stairs
7. Guest Toilet
8. Open Sided Living & Dining
9. Teppanyaki Bale
10. Bedroom
11. Bathroon
12. Coffee Shop & Library
13. Gym
14. Storage
15. Yoga and Meditation Sala
16. Swimming Pool
17. Pool Deck
18. LIving Room
19. Dining Room
20. Bar
21. Kitchen
22. Staff Office
23. Carport
24. Sunset Upstairs Terrace

　非凡的玛拉提娜别墅拥有七间卧室,专为皇室生活打造。别墅位于美丽的传统小村庄乌玛拉斯中,靠近水明漾,四周环绕着水稻,其奢华程度可与高标准的巴厘岛现代寓所相媲美。

　这栋壮观的现代别墅占据了大片土地,汇聚了巴厘岛绝佳度假村的所有优势:七间装饰豪华的卧室;25米长的泳池被日光浴床包围,供你悠闲地消遣时光;图书馆内装有空调,适合早晨在此饮用咖啡或品尝凉爽的鸡尾酒;健身房内装配齐全;瑜伽室可俯瞰美丽的稻田;在铁板烧亭子中等待厨师烹饪能与高级日本餐厅相媲美的美食,富有情趣。

　玛拉提娜别墅拥有一支熟练、专业的队伍,包括别墅经理、管家、服务员、安保人员和室内厨师,他们对你有求必应。不管是庆祝生日,还是周年纪念日,甚至是婚礼,别墅都将给你提供充足的场地和完美的场景,带给你美好的回忆。

　玛拉提娜别墅提供了巴厘岛上最美好、最美丽的东西。在这奢华的别墅中度假,没有人愿意离去。若想体验海滩酒吧、美食餐厅和时尚精品店,繁华的水明漾就在附近,只需十五分钟的车程就能到达。

PRESIDENTIAL VILLA
总统别墅

Contributor:
Banyan Tree Ungasan Bali

The three-bedroom Presidential Villa is a heaven of pleasure fit for royalty, providing guests with stunning views of the Indian Ocean.

The Presidential Villa features an expansive 25-metre infinity pool and garden bale; and indoors, three lavishly designed bedrooms, large living and dining areas, a spa treatment room, a fitness area, and a multi-purpose entertainment and relaxation room.

专为皇家打造的总统别墅内设三间卧室,堪称快乐天堂,可一览印度洋迷人的风景。

别墅外是 25 米长的水池和花园,十分阔气。屋内设有三间宽敞的卧室、大型居住和用餐空间、水疗室、健身房以及多功能娱乐休闲空间。

ABSOLUTE WATERFRONT VILLA IN CAPE YAMU

雅慕角海滨别墅

Contributor:
Hunter Sotheby's International Realty

Location:
Cape Yamu, Phuket

Area:
5,003 m² (Land plot),
1,900 m² (Built-up), 800 m² (Indoor)

This exclusive community of elegant villas, conceived by architect Jean-Michel Gathy and acclaimed designer Philippe Starck, has become synonymous with the last word in stylish oceanfront living on Phuket. The views of the extraordinary seascape that is Phang Nga Bay, together with direct private beach access, make this a truly remarkable setting. Yet the estate is within thirty minutes drive of all the island's golf courses, marinas, shopping centres, restaurants and nightlife.

The sparkling turquoise waters and emerald islands of Phang Nga Bay form a magnificent backdrop whichever part of the villa you are in. Floor-to-ceiling glass doors open onto extravagant terraces and decking, seamlessly integrating the lush views outside with the understated elegance inside. With exquisite cool white interiors by VG21, inspirational architecture by Philippe Starck and Jean-Michel Gathy, gorgeous infinity-edge pool, fitness room, home cinema, spacious terraces and decking, the villa represents a pure fusion of elegance, function and sophistication.

该别墅群别具一格,由建筑师让·米歇尔·加蒂(Jean-Michel Gathy)和备受称赞的设计师菲利普·斯塔克(Philippe Starck)合力打造而成。它是时尚的普吉岛海滨生活的代名词。攀牙海湾的独特海景与完全私人化的海滩入口,成为别墅群的非凡背景。别墅离全岛的高尔夫球场、游艇码头、购物中心、餐厅和夜市只有30分钟的车程。

别墅的任何一处都能欣赏璀璨的蓝绿色海水和攀牙海湾边翠绿的岛屿。落地玻璃门开向宽敞的露台和户外平台,将屋外丰富的风景和屋内低调的优雅巧妙地融合在一起,找不到任何瑕疵。室内用VG21打造成精致的冷白色色调,别墅还设有美丽而宽敞的无边泳池、健身房、家庭影院和开敞的露台与平台,充分展现了其优雅性、功能性和先进性。

BREATHTAKING SEA VIEW VILLA
令人惊叹的海岸别墅

Contributor:
Hunter Sotheby's International Realty

Location:
Kamala, Phuket

Area:
2,300 m² (Land plot), 1,500 m² (Built-up), 650 m² (Indoor)

This luxury five bedroom villa is located within an exclusive private residential development on Phuket's western coast, sitting discreetly on a secluded headland. The project, set at the head of "millionaires row" in Kamala, embraces one of the most beautiful coastlines Phuket has offered.

Constructed mid 2008, this magnificent villa embraces the natural beauty of the surrounding tropical forest and picturesque hills, offering sweeping views of the Andaman Sea, breathtaking sunsets and even capturing the twinkling lights of Patong Bay.

The villa has been meticulously designed, with each room featuring floor to ceiling glass walls and a stunning cathedral-style timber roof which is echoed by the dark walnut timber flooring. Bathrooms feature luxurious marble floors and counter tops, walls and cubicles are bejeweled with metallic mosaic tiles. The latest Vantage Automated Home System enables the home to be transformed from any subdued relaxed mood of your choice, to an uplifting party mode for entertainment at the press of a button. Each room has its own mood lighting and integrated sound system.

Unsurpassed attention to detail has been invested in the landscaping of this luxury estate. The gently sloping plateau that the villa occupies is one of the very few on this site that lends itself to a very family friendly environment. The villa "floats" on three levels, the uppermost level featuring the "Owner's Retreat", whilst the other two levels form a U-shape around a central lily pond at the entrance level, and infinity pool down below, allowing a harmonious flow throughout the rooms and walkways.

http://m.acs.cn/3u4135/

这座豪华别墅坐落在普吉岛西海岸独家私人住宅开发区内，毫不起眼地耸立在一个隐蔽的海岬上，内部设有五间卧室，是卡马拉"百万富翁区"的龙头别墅，拥有普吉岛最美的海岸风光。

别墅建于 2008 年中期，被颇具自然美感的热带雨林和如画般的山岗包围，可远观安达曼海海景、令人赞叹的日落景观和芭东湾的万家灯火。

别墅设计得一丝不苟，玻璃幕墙贯穿每个空间，大教堂式的木屋顶与黑胡桃木地板相呼应，效果非凡。浴室的地面和盥洗台面被大理石覆盖，显得十分奢华。墙壁和隔板则饰以金属色马赛克墙板。最新的自动化家庭设备能使家庭空间转化成任何你想要的休闲气氛，只需按下按钮就可以转换成令人振奋的派对模式，供你娱乐消遣。每个房间都有情境照明灯和音响设备，十分完善。

豪华别墅的景观打造得很细致，无法被超越。别墅矗立在缓坡上，位置优越，不可多得，家庭般的氛围十分和谐。别墅的三层楼仿佛漂浮着，最顶层可谓是"主人的私阁"，另外两层沿入口处的中央莲花池呈 U 形展开，楼下设有无边泳池，使房间和过道贯通流畅，空间和谐而不错乱。

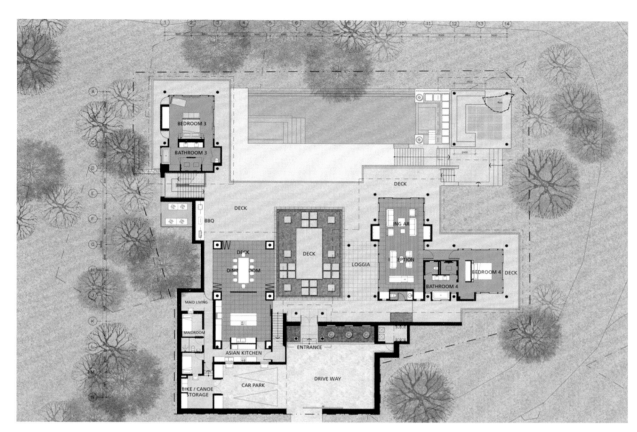

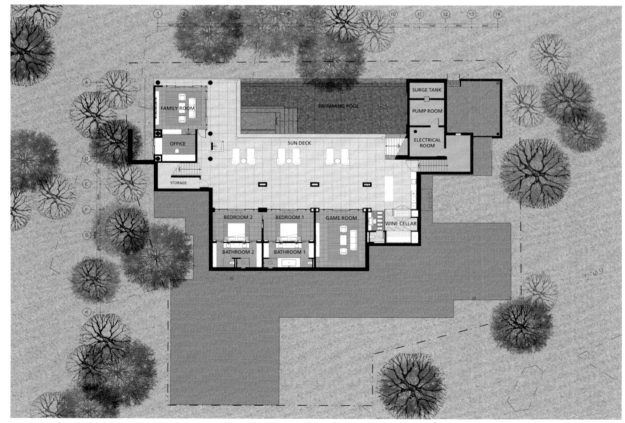

The added advantage is its protection from the harsh south westerly monsoon winds whilst simultaneously capturing welcoming sea breezes year round. Such careful spatial planning also allows all the rooms to enjoy the stunning views of the Andaman Sea. There are numerous areas to relax in: cool shady spots; sunbathing terraces and pavilions, accommodating as much peace and tranquility or socialising as one could hope for, all in complete privacy and harmony with the surroundings.

该别墅的另一个优点是：免受严酷的西南季风侵害，一年四季不断有凉爽的海风吹来。空间结构经过精心规划，使所有的房间都可以享受迷人的安达曼海风光。

当然，此处不乏休闲放松空间：绿树成荫的乘凉处、日光浴露台和凉亭，不管是宁静气氛还是社交氛围，应有尽有，既保护了隐私，又与周围环境和谐统一。

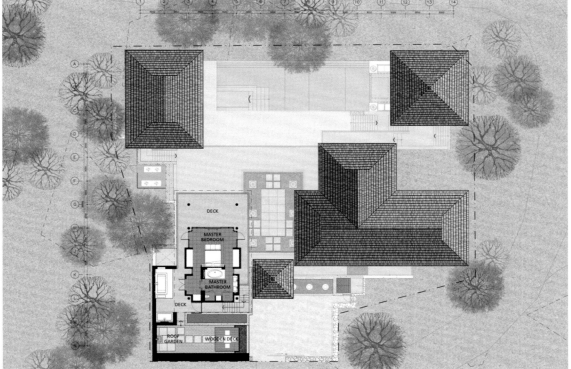

VILLA SAMAKEE
萨马基别墅

Contributor: Awesome Villas	Total Indoor Area: 517 m²	Pool: 100 m²
Location: Phuket, Thailand	Total Covered Outdoor Area: 45 m²	
Total Land Area: 1,846 m²	Total Outdoor Area: 353 m²	

A sense of majesty ensues from the moment you arrive at Villa Samakee. Entering through its oversized teak front doors is to step inside another world; a true tropical escape. A long stretch of glimmering blue meets you on arrival: the 20 by 5-metre infinity pool, where life in the villa revolves.

The spacious wooden pool deck offers a divine choice of spots to catch the sun's rays, while a lovely outdoor sala provides a shaded lounge area just a hop away from the 10-person jacuzzi. Towering fan palms and a rich variety of tropical garden plants create a vivid green landscape and a natural border that keep the villa ultra private and secure.

Inside, a remarkable collection of art and antiques from traditional Thai temple spires to ancient Buddha statues to bright whimsical paintings, including several depicting Asian archetypes by Belgian artist Christian Develter, add character and colour throughout the villa.

A large living room with a gorgeous vaulted timber ceiling and two walls that open completely to the outdoors offers a nice spot for reading or enjoying a cosy chat with the family. The well-equipped Home Theatre is a colourful, fun space for children and adults alike while a private study leading to a garden patio provides a quiet retreat. A formal dining room set in front of the superbly fitted out western kitchen has seating for 16.

Each of the five bedrooms offers a very well appointed, peaceful oasis for couples or kids. Romance abounds in the upstairs master suite, which features a two-person terrazzo bathtub that overlooks the pool from its private bathroom.

The villa manager, chef, maid and concierge service ensure that every need is taken care of efficiently and with utmost professionalism and care. Enjoy a full range of services on offer, from poolside massages to babysitting to booking a round of golf.

初入萨马基别墅，一种庄严感便油然而生。踏过柚木大前门，就来到了另一个世界——真正的热带度假屋。首先映入眼帘的是长长的蓝色水池，水池长20米，宽5米，池水闪闪发光，别墅的生命在此循环。

宽敞的池边甲板提供了多个捕捉日光的绝佳地点；可爱的户外大厅提供了阴凉的休息区，几步外就是能容纳10人的按摩池。高耸的蒲葵和丰富多样的热带花园植物组成了鲜绿的景观，也成为别墅的天然屏障，保证了别墅的隐私和安全。

室内摆放着一系列非凡的艺术品和古董，从传统的泰国尖顶寺庙，到古老的佛陀雕塑，再到鲜艳的奇异画作，无一不令人瞠目。其中包括比利时艺术家德瓦特（也是基督教徒）的几件描绘亚洲原型的画作，给整栋别墅增添了个性和色彩。

　　大客厅的空间上方是华丽的拱形木天花板,两道墙壁都完全向户外空间敞开,是阅读和享受与家人之间惬意聊天的完美场所。装配齐全的家庭影院色彩缤纷,是儿童和大人的娱乐空间;私人书房能通往花园露台,给人以安静空间。商务餐厅设在装备先进的西式厨房前,能够容纳16人。

　　五间卧室都设备齐全,是夫妻和孩子的"宁静绿洲"。楼上的主套房充满浪漫情调,私人浴室中摆放着双人水磨石浴缸,沐浴时还能俯瞰泳池景观。

　　别墅的经理、厨师、女服务员和门房服务用专业的精神和周到的服务来满足每一位客人的要求,既有效率,又能让客人满意。全面的服务项目包括池边按摩、小孩看护,甚至是预订高尔夫球场次。

VILLA KYA

基亚别墅

Contributor:
Unique Retreats

Location:
Koh Samui, Thailand

Photography:
Lesley Fisher

This expansive villa, combines modern luxury with tropical elegance to deliver a sophisticated sanctuary perfect for groups of friends, special events or a peaceful family retreat. Set amongst sprawling lawns and exotic tropical gardens, with a stunning pool oasis, this four-bedroom villa has been designed to encourage guests to relax, unwind and dissolve the stresses of everyday life in a beautiful tropical island setting with captivating views of Bophut, Big Buddha and Samui's sister isle, Koh Phangan.

Villa Kya accords one of those rare once-in-a-lifetime experiences: a dreamy island setting, gorgeous sea vistas and glamorous interiors with beautiful bedroom sanctuaries.

The Master Bedroom is uniquely located with incredible views through the floor to ceiling glass windows, over the pool and sea beyond. Beautiful decor of a truly contemporary Oriental feel with outside private terrace adds to the stand alone features of this room. The well designed large ensuite bathroom creates a romantic environment with its double lovers rain shower and bespoke outdoor circular terrazzo bath with candle lit chandelier.

Each of the 3 guests bedrooms has been lovingly furnished. Accented with colour and mosaic and complemented with detail, creating gorgeous bedroom sanctuaries with breathtaking panoramic vistas as your backdrop. The bedrooms also feature ensuite bathrooms with power rain showers and double sinks.

The stunning infinity pool and pool deck is a standout feature in Villa Kya, with its soft organic shapes and endless views. A small shaded shallow pool is encompassed in the design, ideal for young children, with parents close by on the sundeck, chill lounge or dining sala for peace of mind and ease.

Large open-air living spaces are integral to the design of Villa Kya. Close to the pool deck is the exquisitely designed Chill Lounge offering a unique space for people to kick back and relax on the sumptuous sofas listening to relaxative tunes under the cooling fans enabling guests to freely enjoy the tropical Thai climate, inhale the fragrant smells of the garden and marvel at the breathtaking views in utter peace and privacy. The Dining Sala, also located by the pool deck, is the ultimate dining venue. Refreshing sea breezes keep this area cool and it makes an ideal spot for breakfast, the perfect beginning to any holiday. The Moon Terrace, located close to the living room is an outstanding feature of the villa.

The stunning living room with vaulted ceilings, concertina doors onto the large dining terrace and exquisite bespoke furnishings create a graceful interior offering abundance of space, natural light and another separate place to relax. Guests can enjoy a surround sound movie or surf the Internet on the 1.27 m flat

screen Smart TV whilst reclining on the gorgeous sumptuous bespoke sofa seating.

The well appointed high end kitchen with outside sofa seating terrace drenched in unfolding views, is as always the heart of the home.

  基亚别墅,是现代奢华与热带优雅的结合,给朋友群、特殊活动和家庭度假提供了高端、理想的场所。别墅建在大草坪和颇具异域风情的热带花园中,设有四间卧室,还有令人赞叹的绿洲游泳池,吸引着游客来此放松,在美丽的热带岛屿上消除一天的压力,欣赏博普岛、大佛、苏梅姊妹岛和潘安岛的迷人风光。

  基亚别墅给人一生难求的体验:梦幻的岛屿、壮观的海景、迷人的室内装扮和美观的卧室空间。

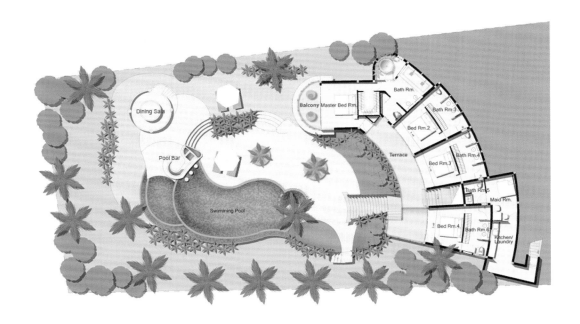

透过主卧室中的玻璃幕墙,从水池到大海,独特的风景令人难以置信。美丽现代的正宗东方风格装饰,户外的私人露台给这个独立空间增添了特色。宽敞的浴室、定制的露天圆形水磨石浴缸,被吊灯的荧荧烛光渲染着,给鸳鸯式沐浴提供了浪漫的气氛。

三间客房的装饰都很温馨。色彩、马赛克和细节的补充都恰到好处,并以全景式的风景为背景,打造出迷人的卧室空间,令人惊叹不已。卧室的配套浴室中还设有电动淋浴设施和两个盥洗台。

惊人的长泳池和泳池露台是基亚别墅的特色。它有着柔和有机的线条,是一道无边无际的景观。被遮蔽的小型浅水池免于太阳照射,适合小孩嬉戏,大人也可坐在附近露台的躺椅上享受日光浴,或在餐厅中寻找心灵的宁静和安逸。

宽敞的开放式生活空间与别墅的设计融为一体。靠近水池的露天平台是设计精

巧的休息厅，人们可以在这个独特的空间中放松休息，坐在奢华的沙发上，听着缓和心情的音乐，吹着凉爽的风，自由地享受泰国的热带气候，花园的花香扑鼻而来，迷人的风景令人赞叹，且享寂静的私人空间。餐厅也设在水池露台旁，是最佳的用餐场所。海风带来了凉意，使这片空间成为享用早餐的理想场所，也是任何假日的完美开端。靠近客厅的赏月平台是别墅的又一特色。

客厅的设计令人惊叹，拱形天花板，宽敞的用餐露台设有六角形门，以及量身定制的室内陈设，显得十分优雅，保证了空间的层次感。空间内自然光充足，还有独立的放松空间。客人可以坐在定制的豪华沙发上，用1.27米高清丽屏平面智能电视观看环绕立体声电影或上网。

高端的厨房装备齐全，坐在户外露台的沙发上，可以欣赏无限风光，这儿便是心灵的家园。

VILLA BAAN WANORA
班瓦诺拉别墅

Contributor:
Unique Retreats

Location:
Koh Samui, Thailand

Photography:
Lesley Fisher

Set along the golden sands of Laem Sor beach, Baan Wanora is distinctly alluring as it embraces its tropical environment, making the most if its 1,579.35 m^2 presence in this stunning beachside locale.

Originally built as a luxurious 5 bedroom family retreat, Baan Wanora has been lovingly designed and furnished to meet every need and wish. This incredibly special residence exudes a vibrant beauty that is a true extension of the family that owns it. Creating a beachside residence with absolute attention to detail.

Stepping inside this private estate, the architectural merit is immediately evident as inside and outside merge seamlessly and beautiful views are a constant backdrop.

The butterfly ponds abundant with lotus flowers and steppers, pave the way to the vaulted living space, center stage in the villa. Large sliding doors concertina back in a unique design dissolving any distinction between indoor and outdoor living.

The sumptuous lounge area has large sofa seating for up to 25 persons, 48" LCD, with an extensive DVD & book library. A simple natural palette spiced with warm vibrant colors and exquisite 100-year-old far eastern antiques, allow this stunning space to speak for itself, as it entices guests to relax and unwind.

Each bedroom sanctuary has been lovingly named after the children in the household. These elegant retreats with sumptuous custom made bedding, furnished with antique pieces and individual color schemes, create gorgeous bedroom havens.

Baan Wanora offers four double bedrooms with stunning views surrounding the pool in this beachside location. A kid's room with 2 single beds and a connecting door to an additional single bedroom, make an ideal children's retreat, close to the main pavilion. The terrace outside features a small plunge pool cooled by an ornamental waterfall. Each room offers highly appointed bathrooms, some with additional outdoor facilities.

Designed with entertaining in mind, gatherings around the large 16-seater dining table are drenched in unfolding views of the villa and beach. The large balcony also features a smaller 8-seater dining table for guests to enjoy some terrace dining. The thoughtfully appointed poolside sala, steps away from the soft white sands and azure waters of the Gulf. Makes an excellent choice for outdoor dining in the lap of luxury, at any time of day.

The scenic seafront pool terrace with palm trees framing perfect vistas of sand and sea follows the dramatic sight lines of the 20m ebony tiled, virtual infinity pool. The impressive salt-water pool is commandingly positioned creating an elegant centerpiece in this well designed space. The swim up pool bar adds an interesting twist to the sleek design of this charismatic pool. Large covered surrounding terraces offer relaxed chill spaces out of the heat of the day whilst bronzed body lovers can enjoy the full glory of the sun on the pool terraces, or the icing sugar sands of Laem Sor beach.

The garden sparkles with the simple elegance of exotic beachside living. Live orchids are attached to each palm tree as indigenous plantings flourish in this lush tropical haven. The scents of the widely planted jasmine and frangipani trees fill the warm evening breezes as kaleidoscopic colors streak the sunset sky.

漫步于蓝索尔海滩的金色沙滩上,班瓦诺拉别墅及其所环抱的热带美景成为这片沙滩上最为耀眼的吸睛之作。这幢占地1579.35平方米的豪华海滨建筑着实令人惊叹。

这座包含了5间卧室的豪华别墅是家庭度假的绝佳选择之一。其充满感情色彩的设计与布置希冀可以满足客人的每一个需求与期望。它特别注重细节,在居住于其中的客人面前绽放出热带活力的美感。

缓缓步入这座私人庄园,室外到室内景观的无缝融合以及其繁花似锦的美景无不体现出其建筑价值。

踏过蝴蝶池中掩映于荷花间的石阶,就能进入别墅的中央舞台——拱顶起居空间。设计独特的可折叠式大型滑动门消融了室内与室外的空间差异。

华丽的休闲区拥有一张足以容纳25人的超大型沙发,配备了48寸的液晶电视、DVD播放器以及私人图书馆。起居区的墙面以简单的色彩打底,适当调入温暖鲜艳的色彩,配以有近100年历史的远东古董,营造出一个使客人得以全身心放松的空间。

别墅的5间卧室都由业主孩子精心命名,让人在感到纯真的同时透露出丝丝趣味。定制的床上用品、古董家具、充满特色的配色方案以及华丽的装潢打造了5间优雅豪华的住所。

　　别墅设有四间双人卧室，室内可以观赏环绕着海滨水池的壮观景色。儿童房内摆放了两张单人床，穿过分隔门便是另外一间单人房，房间靠近主厅，是孩子的理想住处。户外的露台中有小水池，还设有装饰性的瀑布。每间房都配有高档浴室，有的还配有户外设施。

　　出于以娱乐为目的的设计理念，别墅的主体建筑边有一个可供16人同时用餐的池畔用餐凉亭，并且具有极佳的观景条件，别墅主体与海滩一览无余。餐厅还设有可以容纳8人同时用餐的大阳台，映入眼帘的湛蓝海水能够很好地满足野餐爱好者的愿望。凉亭依泳池而建，距海湾的白色沙滩和蔚蓝海水仅几步之遥。在一天中的任何时间，业主们都可以在奢华的户外环境中用餐。

　　别墅的另一大亮点是其精心设计的海滨泳池露台。露台周围环绕着棕榈树，同时可以欣赏到广阔的海景。足有20米长的乌木圈出一座壮美的盐水泳池，其散发着自然气息的同时也透出优雅的内涵。置于泳池内一侧的泳池酒吧为这座大气优雅的泳池带来转折性的魅力。泳池周围覆有大型的露台，为住客带来丝丝清凉。如果你是古铜色皮肤的爱好者，期待体验地道的日光浴，那露台的上方区域以及蓝索尔海滩是很好的去处。

　　傍晚，庄园中充满异国情调的植物与海滨景色交织在一起，透露出古朴典雅的气息；兰花、棕榈树、茉莉以及素馨花共同打造出如万花筒般绚丽的夕阳美景。

VILLA BELLE
贝拉别墅

Contributor:
Unique Retreats

Location:
Koh Samui, Thailand

Photography:
Lesley Fisher

Villa Belle is a spacious and luxurious über styling 3 bedroom villa, set in lush tropical surroundings on a hillside overlooking the captivating bays of Choeng Mon and Plai Laem.

Villa Belle's stunning hillside location, surrounded by lush sloping tropical gardens with impressive feature slate walls and ornamental ponds fringed with traditional Asian carvings, leads guests to the uniquely designed wrought iron main entrance. Arranged over two levels including three spacious luxurious bedrooms, an open plan lavish living area including kitchen, banquet dining and sumptuous lounge, all enveloped in these kaleidoscopic sunrise and sunset sea views.

Once inside, the impressive vaulted ceilings exude wow factor, in this uniquely designed open plan living space, flanked by floor to ceiling disappearing walls of glass, encompassing a fantastic chef — designed gourmet kitchen and breakfast bar, highly appointed with top of the range appliances and accessories.

The large antique dining table has a commanding location drenched in views offering formal dining for up to 8. The marshmallow white sofas and oversized rocking chair ooze the essence of comfort ability with a state of the art LCD TV/DVD entertainment surround sound system encompassing a multi-channel satellite and an Apple TV system to download movies and series, all adding to that perfect holiday ambiance.

Blurring the divisions between indoor and outdoor living, the wrap around expansive wood terrace unfolds to feature an 8 metres Balinese style infinity edge swimming pool, large outdoor lounge sofa seating, with BBQ deck for that ultimate alfresco dining experience.

http://3d.acs.cn/2015/4143/

贝拉别墅高端、阔气且奢华，它拥有三间卧室，坐落在山坡上，四周为茂盛的热带景观，远眺可看见迷人的曾蒙海滩和千手观音庙。

贝拉别墅醒目地耸立在山坡上，被斜坡上茂盛的热带花园环绕，花园又被功能石板墙包围，给人以深刻的印象。装饰性的水池边缘是传统亚洲雕塑，把游客引入到设计独特的锻铁主门。三间豪华卧室分布在两层楼中，开放式的宽敞生活区内设有厨房、宴会厅和休息厅，任何空间都能欣赏到千变万化的日出和日落海景。

步入空间，迷人的拱形天花板便呈现在眼前，令人不禁赞叹。在这个开放式的生活空间中，两侧都是玻璃幕墙，还设计有美食厨房和早餐台，并安装了高端的设施，简直不可思议。

大型的复古餐桌位置摆放恰当，突出了其威武的气质，能满足8人常规用餐。棉花糖般的白色沙发和大型摇椅具有艺术气质，透露出其舒适的本质。空间内设有 LCD TV/DVD 娱乐设施和环绕立体声音响设备，包括多频道卫星电视和苹果电视系统，可以下载电影和电视剧，增添了完美的假日气氛。

广阔的木板露台模糊了室内与室外空间的界限，延伸至8米长的巴厘岛风格的泳池、大型的户外休闲沙发和户外烧烤平台，给人非同寻常的户外就餐体验。

Exquisitely furnished throughout in an elegant contemporary Asian style, fused with antique accessories of the Orient, reflecting a luxury tropical island life style. Soft colours of white and cobalt blue blend with the beautiful coastal surroundings mingled with soft black accents creating an interior designers dream villa, with impeccable attention to design detail.

The Master suite is indeed befitted with all the grandeur it deserves. Walls of disappearing glass overlook the Choeng Mon and Plai Laem bays and large wood decking that wraps around the villa.

White wood floors contrast with the impressive King Size black surround bed and the stunning contemporary black chandelier holds court, centre stage in this outstanding Master Suite. Black and white accessories adorn the shelves and well-captured photos of the island add a homely feel with their juxtaposition above the master bed.

The 2 guest bedrooms are magnificently designed on a floor of their own. Spacious rooms with exquisite four-post beds reside commandingly, affording each bedroom the captivating views that flank these designer sanctuaries from every perspective. Oversized windows emanate grandeur and present the gorgeous outdoor chill out lounges exclusive to these suites, back dropped by the beautiful unfolding island views. Fitted with individual entertainment systems, the bedrooms have been furnished with sofa seating so as to enable guests to enjoy some of their own quiet time should they desire.

优雅的当代亚洲风格设计中掺入了精致的家具和东方复古饰品，反映出奢华的热带岛屿生活。柔和的白色和钴蓝色与美丽的海岸风景融为一体，柔和的黑色调打造出了室内设计师梦想中的别墅，每个设计细节都无可挑剔。

主套房内的装饰的确称得上是豪华壮观。玻璃幕墙外是包围别墅的木板露台，可以远眺曾蒙海滩和千手观音庙边的海滩。

大号床的床托和当代风格的吊灯都为黑色，与白色的地板形成对比，成为主卧室套房的焦点，震撼人心。黑、白色的饰品点缀着书架，床上方的一排照片捕捉到了岛屿的美景，给人以家的感觉。

两间客房在同一层，其设计十分壮观。精致的四柱大床威严地耸立在宽敞的卧室中，于空间的任何角落都能欣赏到无限的风光。大型的窗户十分宏伟，华丽的户外休息空间是这一空间的特色，无限延伸的岛屿美景也成了该空间的幕景。卧室内配有个人娱乐设施，客人也可以坐在沙发上享受自己渴望的安静时光。

VILLA KALYANA
卡雅拿别墅

Contributor:
Unique Retreats

Location:
Koh Samui, Thailand

Photography:
Lesley Fisher

The vaulted pyramided lounge pavilion holds a commanding presence at the top of the estate. The outstanding design encompasses the breathtaking views and impressive beach frontage, in unprecedented luxury.

The villa central focal point is the impressive 300 m^2 chemical free pool that resides centre stage amidst the tropical palms. The expansive wooden and marble pool decks offer loungers and sun beds, ideal for sun worshipers. Hammock Island, the central pool feature, is perfectly positioned for those wishing to relax.

Working on the premise that everyone deserves a little luxury, Villa Kalyana is here to deliver quite simply that taste of luxury, steeped in beauty island style.

The surrounding gardens and pool decks are vast in size and gorgeous by natural design. The large lawn peppered with palms, offers guests a diversity of spaces to sunbathe, relax or simply play games in this tropical poolside haven.

Most if not all coconut trees have been preserved to maintain the local insight and natural wonder of this exceptional isolated place. Outdoor terraces are styled with native plantings to encourage privacy.

Each of the fourteen beachfront and sea view bedroom are drenched in unfolding views with the utmost attention to detail. The suites are highly appointed, with Smart Internet LED TV's and Ipad / Ipod Docking Stations. Individual interior designs feature exquisite far eastern antiques, enhancing the Orient feel of the estate.

http://3d.acs.cn/2015/4142/

The home theater system, with its 55″ HD TV and surround sound, creates a great movie room. Sumptuous contemporary sofas are ideally positioned for perfectly chilled movie viewing at any time of day.

The beauty of this space is its incredible flexibility in transforming itself to one's need.

卡雅拿别墅顶上的拱形金字塔式凉亭姿态特别，格外引人注目。杰出的设计充分融入了令人赞叹的风景和引人注目的滨海海滩风光，令别墅极其奢华。

别墅的焦点是热带棕榈树中间的 300 m² 的无化学物质水池，令人印象深刻。水池的周围用木材和大理石覆盖，并摆放着躺椅和日光浴床，是太阳崇拜者们的理想场所。"吊床岛"是水池的特色，是渴望放松的人们的完美场所。

在每个人都应得一丝奢华的前提下，卡雅拿别墅传达出极其简单的奢华感，融入到美丽的岛屿风格中。

环绕别墅的宽阔花园和池边平台因自然的设计而更加美丽。宽广的草坪上种着棕榈，给客人提供了多样的空间，让人们在这热带水池边享受日光浴、放松和嬉戏。

为了维持这片单个场地的风景和自然景观，保留了大部分的椰子树。户外露台则用当地植物来保护隐私。

十四间朝向海滩和海洋的卧室都沉浸在广阔的风景中，细节之处极具吸引力。套房中设备先进齐全，比如智能网络、LED 电视和 Ipad、Ipod 接口。个性化室内设计的特色是精致的远东古董，增添了空间的东方气质。

宽敞的电影间设有家庭影院系统，包括 55 英寸高清电视和环绕立体声音响。华丽的现代沙发恰到好处，能满足任何时间观看电影的需求。

该空间的美在于令人难以置信的灵活性，能根据人的需求随时转化。

The large size of this room and floor to ceiling windows over the garden and pool make for a beautiful backdrop when the area is transformed to a conference room. The villa has a large desk with conference style seating that can be set up. The space is also suitable for small presentations and business meetings with supported business services, including printing, copying and faxing.

The villa's Games Area is a fantastic addition to this property, with a full size snooker table, on the lower level for snooker enthusiasts to keep amused.

宽敞的房间和玻璃幕墙朝向花园和水池，是会议室的美丽背景。别墅内设有大型会议桌和会议风格的座椅、商务服务设施，如打印、复印和传真设施，使空间同样适合小型展示和商务会议的要求。

别墅的游戏区是房子的奇妙一笔，宽大的台球桌能使斯诺克爱好者玩得尽兴。

Ideally situated with a unique panoramic view of the lush tropical gardens and mosaic tiled azure pool and golden sandy beach, the gym at Villa Kalyana has a variety of high- tech fitness equipment, in air-conditioned luxury. Remarkable panoramic vistas viewed through floor to ceiling windows evoke a feeling of wellbeing for exercising the body with a rewarding cool down swim just a few steps away.

别墅占据了优越的地理位置,能观赏到茂盛的热带花园、蓝色马赛克水池和金色的沙滩,所有景色一览无遗。内部的健身房十分奢华,拥有各种高科技健身设备,还装有大型空调系统。透过玻璃幕墙便是狭长的全景式景观,给人以健身的幸福感,而且,只需几步就能享受舒适的游泳运动。

VILLA LEELAVADEE
里拉瓦迪别墅

Contributor:
Unique Retreats

Location:
Phuket, Thailand

Photography:
Lesley Fisher

Villa Leelavadee is a holiday home for the discerning world traveler that cleverly brings together the traditional with the contemporary. Built as a celebration of island sophistication, soft decorative wood accents and plush furnishings conjure up a genteel, homey ambience that beautifully balances the villa's modern character.

A breathtaking ocean panorama dramatically unfurls as guests descend from the villa's elevated entrance into the main lounge with views that take in a long crescent bay and extend out into the azure waters of the Andaman Sea. Making the most of the stunning tropical surroundings, a covered outdoor terrace has been added to the main living room area, allowing the villa's indoor space to segue seamlessly outdoors with the balmy ocean breeze drawn in through folding glass doors.

The infinity pool glitters under the tropical sun, while a poolside relaxation pavilion provides shaded reprieve and an opportunity to soak in the beauty and tranquility of the location. A private path meanders down the side of the hill directly from the villa grounds, down to a secret beach, adding an extra layer of privileged exclusivity to any stay at Villa Leelavadee.

Intelligent design features include glass panels in the swimming pool that cast waves of sunlight onto the level below, bathing the games area with natural sparkles. A private gym, equipped with fitness machines and free weights, plus a personal sauna, ensure guests never have to leave the villa's captivating surrounds to stay in shape. To complement the indoor facilities, a private 13-meter pool provides an opportunity to stretch and breathe in the clean island air while also enjoying the tropical vistas.

http://3d.acs.cn/2015/4149/

LEELAVADEE - Villa 6
GARDEN LEVEL
Approx. 533 sq.m.

NOTES:
- Single beds in Bedrooms 3 & 4 may be combined to form Kings

LEELAVADEE - Villa 6
POOL LEVEL
Approx. 545 sq.m.

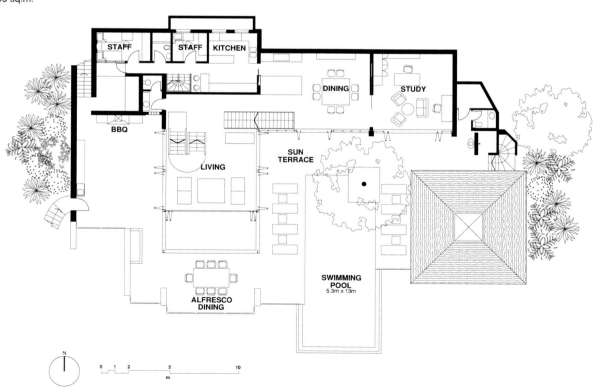

里拉瓦迪别墅,是独具慧眼的游客选择的假日家园,巧妙地融合了传统与当代元素。它具有鲜明的海岛别墅风格,屡次使用色彩柔和的装饰性木材,与豪华的陈设一起营造了文雅而亲切的氛围,与别墅的现代感毫不违和而更显相得益彰。

从别墅的入口进入到主客厅,从新月形的海滩到蔚蓝的安达曼海,海洋全貌风光在眼前铺开,引人注目,令人赞叹。为了充分利用引人入胜的热带环境,给主客厅空间增加了户外凉亭,温暖的海风从折叠玻璃门吹进来,把别墅的室内空间与室外空间融入在一起。

长长的水池在热带阳光下闪闪发光,旁边的休息亭是清凉的慰藉,这儿的美丽和静谧令人沉醉。山上的羊肠小道直接从别墅延伸至秘密海滩,给里拉瓦迪别墅增添了一层特殊的独有性。

泳池的玻璃墙设计巧妙,把阳光导入到下层,游戏室就像沐浴在自然光中。私人健身房不仅有健身器械和力量训练器械,还有单人桑拿浴,客人无需离开别墅就能健身。13米长的私人泳池丰富了室内设施,客人不仅能在岛屿中伸展自我,呼吸新鲜空气,还能欣赏热带远景。

ICONIC ESTATE
雅致房产

Contributor:
Hunter Sotheby's International Realty

Location:
Nai Thon, Phuket

Area:
6,983 m² (Land plot), 1,218 m² (Indoor), 374 m² (Terrace)

First time to market, this magnificent oceanfront resort property is the ultimate sophisticated waterfront abode and sits on nearly 7,000 sqm of land, with unsurpassed panoramic views of the Andaman Sea and stunning sunsets. Located within the private villa estate at Trisara Resort, the expansive and elegant six suite villa features highly crafted detailing and impeccable finishes throughout.

This iconic property includes a private café, spa with steam room and sauna, home theatre, fully equipped gymnasium and two infinity-edge swimming pools, and is immaculately appointed and maintained to the highest of standards for a truly unique lifestyle.

This extraordinary home is positioned within lushly landscaped tropical gardens and manicured lawns which stretch down to the ocean's edge. Furnished with unique original artworks and antiquities, this is one of the most sought after properties in Southeast Asia.

http://m.acs.cn/3u4134/

这栋初次进入市场的海滨房产非同寻常，完全坐落在海滨位置，占地面积 7000 平方米，能欣赏到无与伦比的全景式安达曼海风光以及迷人的日落风光。该栋优雅的建筑处于私人别墅房产特瑞萨拉度假村内，内部设计了六间套房，细节和内饰都经过了精心雕琢，无可挑剔。

该标志性建筑内设有私人咖啡厅、水疗房（包括蒸汽房和桑拿房）、家庭影院、装备齐全的健身房和两个无边无际的泳池，完美胜任并满足了高标准的独特生活方式。

茂盛的热带景观花园和修剪整齐的草坪环绕着非凡的住宅，室内用别致的原创艺术品和古董装饰着，这是东南亚最受欢迎的别墅之一。

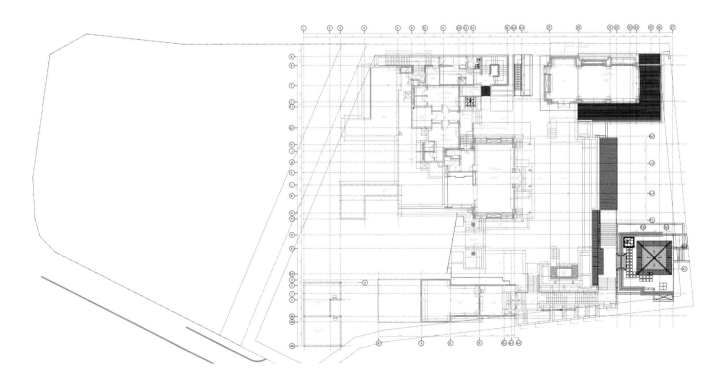

VILLA SHAMBALA
香巴拉别墅

Contributor:
Marketing Villas Ltd.

Location:
Central Seminyak, Bali

Villa Shambala is a uniquely spacious property in the heart of Seminyak boasting an eclectic fusion of old and new, borrowed largely from the traditional wooden joglo houses of Java, but with a stylish blend of contemporary comforts.

Set on a large estate, the single-story building comprises a main pavilion and a separate guesthouse, set at opposite ends of the dramatic 19m swimming pool. Villa Shambala boasts the distinctive trapezium-shaped roofs of the antique joglo style, but does away with the enclosed, dark walls of this conventional design in favour of floor-to-ceiling windows and folding glass doors blurring the lines of indoor-outdoor space and offering guests a choice of alfresco or enclosed living. The open-plan design of the main building allows beautiful natural light into the central living and dining areas and accentuates the honey-coloured detail of both recycled teak and ornate carvings that are the signature features of Shambala. An ornate chandelier decorates the main living area sitting poised above a plush U-shaped lounge for the ultimate in relaxation.

The living area spills onto a large alfresco dining area which can be fully enclosed and air-conditioned, or open to the breeze. The expansive full-service kitchen is a key feature of Villa Shambala giving guests the option of cooking for themselves, or utilising the villa's private chef and service staff. The modern, communal kitchen is integrated into the main living area – perfect for dinner parties, socialising and private events.

　　香巴拉别墅坐落在水明漾中央，新和旧的折衷融合风格使建筑独特又宽敞，十分突出。别墅大力借鉴了爪哇岛传统的佳格洛木屋风格，但又结合了现代的舒适性，十分时尚。

　　别墅只有一层，但占地面积广，设有主楼、独立客房和另一端的 19 米长的大型泳池。香巴拉别墅因其特有的梯形屋顶而引人注目，具有古老的佳格洛屋风格，但去掉了传统设计中的封闭式深色墙壁，而采用了玻璃幕墙和折叠玻璃门，从而模糊了室内与室外空间的界线，给客人提供了户外生活或封闭生活两种选择。主建筑结构的开放式设计让明媚的阳光深入到中央的客厅和餐厅，也使蜜色的再生柚木和装饰性雕刻品更加醒目，这就是香巴拉别墅的主要特色。华丽的大吊灯装饰着主生活区，下方的 U 形毛绒大沙发给人极致的舒适感。

客厅面向宽敞的露天用餐区，此空间既能封闭起来，用空调调节气温，又能直接敞开，让清风吹入。宽敞的厨房也是香巴拉别墅的一个重要特色，能提供全方位的服务，住客可以自己烹饪，也可以享受别墅的私人厨师和服务人员提供的服务。此现代化公共厨房与主生活区融为一体，是宴会、社交和私人活动的绝佳场所。

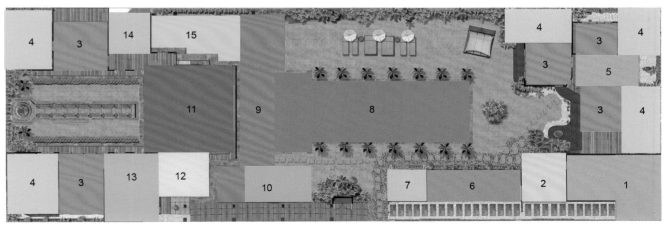

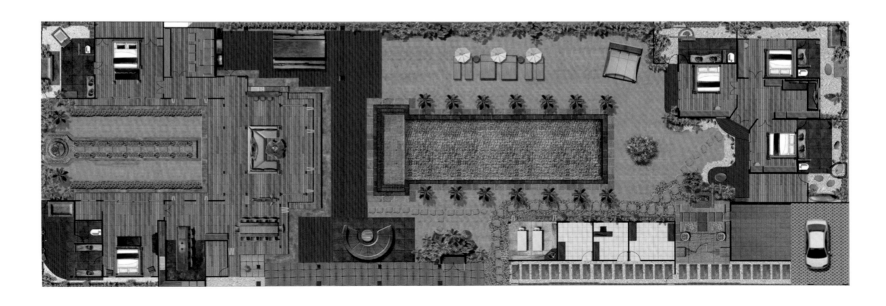

The master bedroom features a hand-carved teak joglo-style canopy, with huge outdoor ensuite, open-air stone tub, separate shower and double vanity. The master and second bedroom sit beyond the living areas, to the rear of the villa, and both overlook a peaceful water feature complete with large stone Buddha and tranquil lily pond. Villa Shambala features a fully equipped media room complete with large screen projector, surround sound system, DVD and satellite TV with an extensive library of DVDs, CDs and books for guests' enjoyment. Wifi is also available throughout the villa.

The living room incorporates the hand-carved teakwood core of an antique joglo, while timeworn statues from the eastern isles have been fondly placed in well-lit settings along with various beautiful pieces from all over the world — a testament to the owner's travels. Leading away from the main building, and pasting a circular, shaded outdoor lounge, sit a private double spa room where guests can enjoy massage and Ayurvedic treatments administered by skilled therapists. Beyond the pool sits the second pavilion, which houses three queen-sized rooms, surrounding a common living area which is fully equipped with an entertainment centre — perfect for children. Each of these queen bedroom features its own outdoor ensuite bathroom and the beds can be configured as queen or twins for guest flexibility.

主卧室中设有手工雕刻的佳格洛风格柚木华盖，还配有大型户外石浴缸、单人浴室和双人盥洗台。主卧室和第二卧室都设在主生活区上方，即别墅的后部，可远眺静谧的水景，大石佛陀和平静的莲池使空间更加完善。香巴拉别墅还设有媒体室，室内装备齐全，包括大型投影仪、环绕立体声音响系统、DVD机和卫星电视，除此之外，还有大量的DVD、CD和书籍可供客人使用。无线网络覆盖了整栋别墅。

客厅以古佳格洛屋的核心——柚木雕刻的手工为基调，来自东部群岛的陈旧雕像坐落在灯光之中，惹人怜爱，空间还摆放了来自世界各地的各种美丽饰品，纪念了房主的每次旅行。步入主建筑，穿过环形的带顶户外休息厅，便来到了私人的双人水疗室，住客能够享受按摩和专业理疗师的阿育吠陀疗法。水池的另一边是第二栋建筑，内部设有三间大房和公共生活区，后者还带有装配齐全的娱乐中心，是孩子们的最爱。每间大卧室都带有独立的露天浴室，床可以根据客人的需求组合成大床或双人床。

ESHARA VILLAS
伊莎拉别墅

Contributor:
Marketing Villas Ltd.

Location:
Seminyak, Bali

Situated off a quiet street, yet only 300 metres from Seminyak's bustling beach, shops and restaurants, Eshara villas I, II and III each have their own private gardens and swimming pools — separated by a clever system of retracting garden walls — as well as individual indoor and outdoor living spaces and equipped, open-plan kitchens. When the whole villa is opened up as one, it can easily cater for up to 16 guests.

The indubitable sexiness of our interior design is intended to delight: think funky furnishings, brightly coloured artwork, fascinating artefacts and lush fabrics; a place where stone-carved Balinese goddesses invite you to be seated on cowhide chairs under opulent Swarovski crystal chandeliers. And this tropical living concept spills outside onto palm-tree-dotted lawns where you can lounge on pool decks or enjoy romantic dinners lit by flickering fire bowls.

Our team of highly competent and professional staff will make your holiday wishes come true. The villa manager and personal butlers are on hand to ensure you want for nothing and, if you'd rather not self-cater, our chef will prepare your meals from a large suggestion menu of Asian and Western favourites or according to your special requests.

With a staggering array of hedonistic pleasures virtually on the doorstep, Villa Eshara's location is a wonderland for bon-vivants. From designer boutiques and day spas to beach bars, world-class restaurants, pulsating nightspots and the expansive Petitenget Beach, fun and fine-dining are never much more than a few moments away.

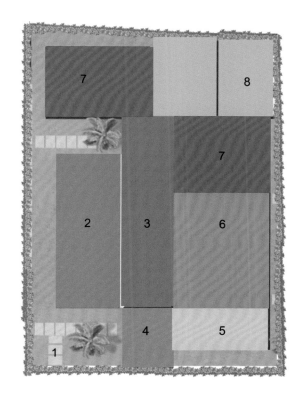

KEY PLAN

2-Bedroom Villa Eshara II

1. Villa Entrance
2. Swimming Pool
3. Pool Deck
4. Staff Area
5. Kitchen
6. Living & Dining Area
7. Bedroom
8. Bathroom

伊莎拉别墅坐落在安静的街道上，离热闹的水明漾海滩、商店和餐馆只有300米。 三栋伊莎拉别墅都各自设有私人庭院和泳池，且被花园墙巧妙地隔离开来。房子还包括室内、室外私人空间和设施齐全的开放式厨房。 当整栋别墅被敞开时，令空间融为一体，能够轻松容纳16人。

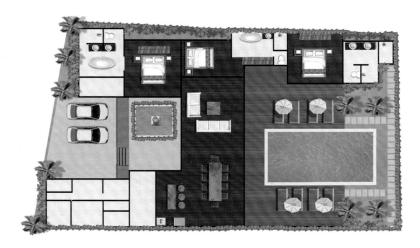

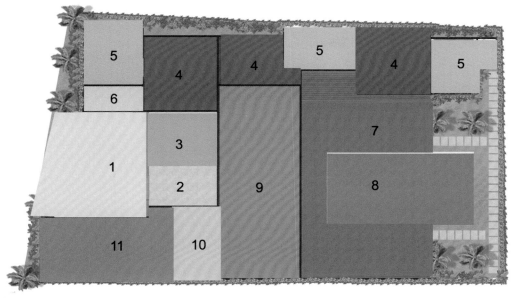

KEY PLAN
3-Bedroom Villa Eshara I

1. Parking Area
2. Villa Entrance
3. Water Feature
4. Bedroom
5. Bathroom
6. Security Post
7. Sun Loungers
8. Swimming Pool
9. Living & Dining Area
10. Kitchen
11. Staff Area

室内设计的魅力无可置疑，旨在愉悦心情：时尚的陈设、色彩鲜艳的艺术品、迷人的手工艺品和丰富的织物。巴厘女神石雕像吸引人们进入空间，坐在牛皮椅上，头顶便是施华洛世奇水晶吊灯。户外的草坪上点缀着棕榈树，沿用了热带生活理念。在池边露台上休息，或是在烤炉的摇曳火光下享受浪漫的晚餐，别有一番风味。

专业能干的团队将帮你实现假日愿望，别墅经理和私人管家随叫随到，服务周到。如果你不想自己动手烹饪，主厨将给你准备美食，菜单上的亚洲和西式美食，随你挑选。

伊莎拉别墅就像是享乐主义者的乐园，当人们踏上台阶，就能感受到那种欢愉。从时髦的精品店到日间温泉，再到沙滩酒吧、世界顶级餐厅、繁华的夜总会和广阔的佩特坦盖特沙滩，娱乐和就餐地都在几分钟的距离内，十分便利。

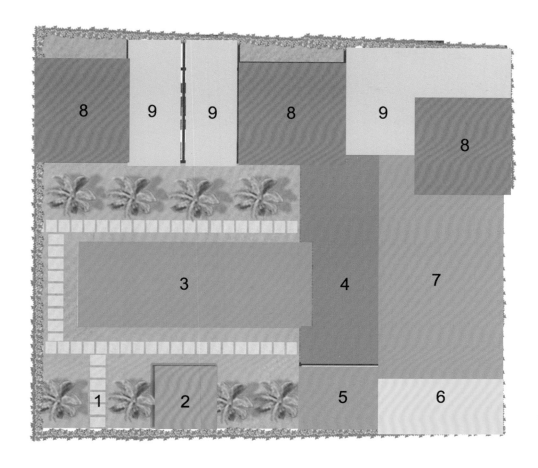

KEY PLAN
3-Bedroom Villa Eshara III

1. Villa Entrance
2. Bale
3. Swimming Pool
4. Pool Deck
5. Staff Area
6. Kitchen
7. Living & Dining Area
8. Bedroom
9. Bathroom

VILLA LIBERTY
自由别墅

Contributor:
Awesome Villas

Location:
Phuket, Thailand

Villa Liberty is an awe-inspiring, luxury, six bedroom pool villa situated on the Kamala headland — one of Phuket's most exclusive addresses on the famed "Millionaires Mile". The villa rests on five separate ocean facing levels that are joined by a series of elevators and marble staircases. This designer villa also features a private man-made beach, making it one of the most unique villas in Asia, the very thing that beautiful holiday memories are made of and a veritable heaven of luxury for the discerning visitor.

The Modern Asian-style architecture blends seamlessly with the functional elements of Western design to present a superior, contemporary tropical abode, perfect for any dream vacation for families or groups. With six beautiful bedrooms, a huge cinema for those nights in and a vast entertainment room that lends itself to parties and get-togethers on a whole new level.

The master European and Thai kitchen is a gourmet enthusiasts dream and features all the latest amenities and kitchen appliances thus ensuring the chef can prepare any sumptuous meal that you so desire whilst you enjoy not only a splendid holiday, but a feast that will endure in memory far beyond your time with us. The glorious sparkling pool has its very own bar and barbeque area for you to enjoy a beverage or a lazy afternoon amidst a perfect outdoor setting. A private sala overlooking the Andaman Sea is a dreamy heaven of breathtaking views, postcard perfect vistas and a beacon of peaceful tranquility. Without doubt Villa Liberty offers the ultimate in luxury vacation rental lifestyle and continues to amaze even the most ardent of luxury travellers.

The stunning living area comprises of an astonishing 1,213 square meters (13,000 square feet) of pure vacation luxury and includes a large crystalline pool on the meticulously landscaped primary living terrace with a stunning architectural entrance. The finishes are of the highest grade of terrazzo and teak and are framed with breathtaking ocean vistas. A blue Coral Beach lies below and the glittering lights of Patong can be seen in the panoramic views to the south.

Liberty is an exemplary vacation home where your every need and desire will been taken care of, and exceeded passionate.

http://m.acs.cn/3u4120/

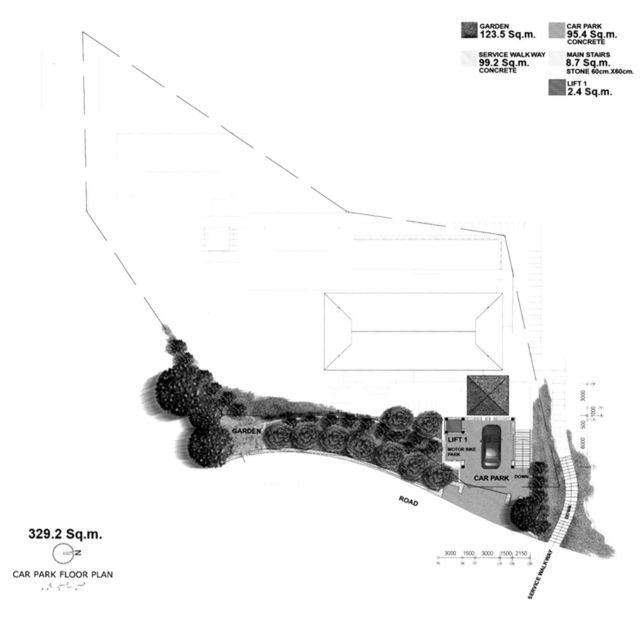

　　自由别墅坐落在著名的普吉岛百万富翁专属区——卡马拉海岬上，它豪华奢靡得令人敬畏，内设六间卧室，还带有水池。别墅设有独立的五层楼，每层都面向海洋，通过电梯和大理石楼梯相互连接。别墅设计独特，还带有私人人工沙滩，这是假日回忆中不可缺少的一部分，也使其成为亚洲最独特的别墅之一，是具有慧眼的游客名副其实的奢华天堂。

　　现代亚洲风格的建筑中融合了西方设计的功能元素，展现了一处高端、现代化的热带居所，是家庭和团体度假的理想场所。别墅内设六间美丽的卧室，宽敞的电影间适合夜间消遣，大型娱乐间把派对和聚会推向全新的层面。

　　主厨房融合了欧式和泰式两种风格，是美食爱好者的梦想，各种最新的设施和厨具可供厨师烹饪任何你渴望的豪华美食。在此，你不仅能享受到一个辉煌的假日，随着时间的流逝，还能让你的记忆回味无穷。波光粼粼的水池边设有吧台和烧烤区，让你在完美的户外设施中享用饮料，享受懒洋洋的下午。私人大厅拥有安达曼海景，是欣赏迷人美景的理想天堂，给人完美的景致和宁静。毋庸置疑，自由别墅提供了奢华至极的度假租住生活，最奢侈的旅客也为此赞叹不已。

　　宽敞的起居区面积为1 213平方米（13 000平方英尺），拥有纯粹的假日奢华。清澈见底的大水池被精心美化的主要生活露台包围，建筑的入口也令人震撼。高级水磨石、柚木都被采用到装饰上来。空间还拥有壮观的海洋远景，下方是蔚蓝的珊瑚滩，面向南面，便能远眺芭东的万家灯火。

　　自由别墅是度假别墅的典范，其服务周到热情，能满足你的所有需求，帮你实现愿望。

VILLA PADMA
帕德玛别墅

Contributor:
Awesome Villas

Location:
Phuket, Thailand

Total land plot area:
4,464 m²

Total covered area:
800 m²

Total uncovered area:
290 m²

Total lawn and garden area:
596 m²

Experience paradise at Villa Padma. Settled seafront on the east side of Phuket, guests at Villa Padma will be entranced by the display of colors from the lush greenery of the mangrove forest to a picturesque evening sunset. Villa Padma features four extravagantly sized bedrooms positioned privately amongst unique living areas to enjoy and share with friends and family.

Beautifully hand-carved traditional Thai friezes and sculptures, as well as more contemporary artwork, adorn the spacious villa amidst well-appointed furnishings.

Villa Padma treats its guests to the highest form of luxury. Guests can enjoy a delightful scene on any afternoon, immersed in the inviting 25-meter pool with submersed beds or sun bathing on the spacious pool deck with luxury lounge chairs. An evening cocktail can be shared in the water fringed sunken sala before dining under the uniquely luminescent ox bone chandeliers in the vaulted foreroom. Challenge a friend to a game of Pool or Football in the games room while others relax in the cinema room featuring a pixel-perfect projection screen with surround sound. The poolside parlor at the heart of the villa is the perfect rendezvous for guests to exchange tales of their day's adventures. When alone, the serene vista from this room had many guests transfixed, surrendering themselves to the calm and tranquility only nature can bestow.

http://m.acs.cn/3u4121/

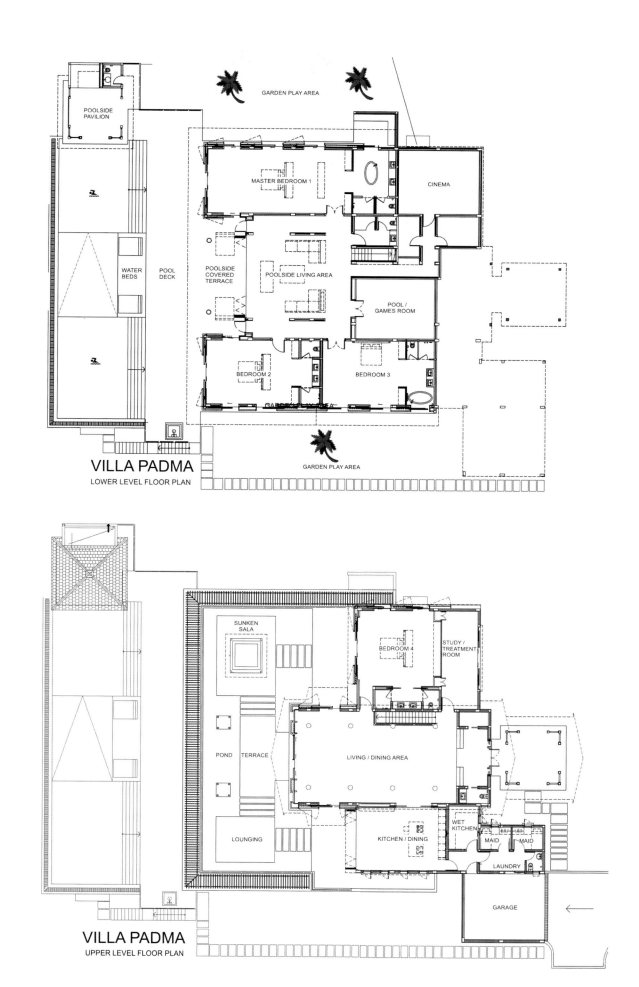

VILLA PADMA
LOWER LEVEL FLOOR PLAN

VILLA PADMA
UPPER LEVEL FLOOR PLAN

Each bedroom includes a desk with a multitude of international sockets as well as WIFI for any unfinished work. A leisurely shower or soak is enhanced by the vast rain showerheads in the bathrooms or the beautiful Terrazzo soaking tubs. The staccato of fishing boats drifts away when the Sonas integrated stereo system plays your music of choice. While relaxing in the bedroom, enjoy the sound of your personal playlist using the specialized Yamaha iPod dock.

帕德玛别墅,体验乐园之趣味。别墅坐落在普吉岛的东侧海滨,红树林的葱郁,如画的日落风景,各种色彩交织出一场视觉盛宴,令人沉醉。内部设有四间独立的超大型卧室,中央是起居区,适合朋友、家庭共同使用。

宽敞的空间内家具齐全,还有各种装饰品,包括传统手工雕刻的精美泰国雕带、雕塑,以及许多当代艺术品。

帕德玛别墅极尽奢华地款待顾客。客人可以在任何一个下午享受愉悦人心的景色,沉醉于25米水池的水上躺床中,或是在池边甲板的躺椅上享受日光浴。夜晚可以在凹陷的水帘大厅中举行鸡尾酒舞会,接下来也可以在拱顶客厅的牛骨吊灯的独特灯光下享用晚餐。在游戏室中与朋友进行台球或桌上足球对决也未尝不可;其他人可以在电影室的高清银幕和环绕立体声音响设备中放松自我。别墅中央的池边客厅是客人聚会的最佳地点,可在此分享他们一天的经历。独自一人在屋内欣赏宁静的远景,在唯有自然能赋予的宁静中放纵自我,也是许多客人梦寐以求的场景。

Fitness on site in the Cape Yamu development is accessible with two tennis courts and ample room for running on the private beach. Breeze Restaurant, featuring Thai and western fare, is located on-site overlooking the water. Lounge on nearby beaches, with Surin, Bang Tao and Layan only 25 minutes away.

每间卧室摆放了一张装有多孔国际插座的桌子,无线网能帮助你完成任何未完成的工作。浴室中的大型淋浴喷头和水磨石浴缸给人放松的淋浴或泡澡体验。当Sonas 集成立体声系统播放着你选择的音乐,渔船的喧嚣声便悄然消失。听着雅马哈平板中个人播放列表中的音乐,在卧室中放松自我也不乏乐趣。

雅幕角的健身设施也可供客人使用,包括两个网球场和宽敞的私人沙滩跑步空间。布雷兹餐厅也在此地,以泰式和西式美食为特色,能远眺海景。若想在沙滩上放松休息,25 分钟就可以到达附近的苏林、邦陶和拉扬海滩。

VILLA RAK TAWAN

瑞克塔湾别墅

Contributor:
Awesome Villas

Area:
3,000 m² (Plot size), 1,800 m² (Villa)

Relish in the luxury design of the three-level Villa Rak Tawan with its stunning views and smart layout which provide maximum privacy for the entire complex and each of the villa's six bedrooms.

Enter your own tropical wonderland in an estate built with soothing natural materials like local stone and teak wood. Bask in the luxury of contemporary tropical design that blends Chinese furniture and paintings with the comfort of Western divans and electronics.

The Upper Level acts as the villa's entertainment and living areas. This level of Villa Rak Tawan features Phuket's best collection of Asian themed antiques and paintings. The Western-style kitchen, with fixtures by Gaggenau, features all the desirable conveniences found in the best European homes. A large, elegant dining room set conveniently nearby can comfortably seat ten guests. Privately hidden, you will find an elegant and secluded bedroom featuring a luxurious balcony overlooking the sea.

The Middle Level is a heaven for entertaining and relaxing, designed around the decadent 16-meter long infinity pool. The two bedrooms on this level are set apart at opposite ends of the pool, each with its own shaded area for private romancing and relaxation. Also on this level you'll find a fully stocked cocktail bar between two separate areas with extensive lounge seating and a large table suitable for the ultimate experience in outdoor dining.

The Lower Level includes two main bedrooms ideal for couples who appreciate their privacy. It offers the same amenities and layout as the Middle Level. The office is spacious and inviting with lovely sea views. It has been designed for quiet studies or family relaxing, and offers a sofa settee with pullout bed perfect for the younger members of the family. The level also is highlighted by a fully equipped professional gym with its own sound system.

瑞克塔湾别墅分为三层，奢华的设计别有一番风味，迷人的风景令人沉醉。巧妙的布局实现了整栋别墅及六间卧室的隐私最大化。

步入这一热带仙境，当地的石头和柚木等自然材料触目皆是，令人舒心。现代的设计中融入了中式家具和画作，再配上西式沙发和电器，奢华而不乏舒适。

顶层是别墅的娱乐区和起居区，内部饰有普吉岛上最好的藏品——以亚洲为主题的古董和画作。西式厨房中配有嘉格纳厨具，具有最好的欧洲厨房所拥有的便利，合乎人意。旁边是宽敞优雅的餐厅，十分方便，能够轻松容纳十位客人，舒适而不拥挤。卧室十分隐蔽，却也不乏优雅，豪华的阳台拥有海上远景。

中间层围绕 16 米长的水池而建，是娱乐和放松的天堂。两间卧室分别设在泳池的两端，每间都带有遮阳顶，满足私人的浪漫追求和放松。该层的两片独立区域中间还设有摆满鸡尾酒的吧台，以及阔气的休息座椅和大桌子，给人极致的户外用餐体验。

底层设有两间主卧室，最适合寻求两人空间的情侣。该层与中间层有着同样的设施和布局。办公室十分宽敞，拥有令人着迷的海景，适合安静地读书或和家人一起放松，其间摆放的靠背沙发可以转变成床，适合家庭中年龄较小的成员。该层的另一亮点是装配齐全的专业健身房，音响设备使空间更加完善。

BULGARI VILLA
宝格丽别墅

Contributor:
Firefly Collection

Location:
Bali, Indonesia

Area:
1,300 m²

Think high-end Italian glamour. Think Balinese spiritual beauty. Blend them together and this is the result. The Bulgari Villa is a dazzling jewel adorning a Bali cliff top, part of the exclusive Firefly Collection of luxury villas from around the world.

Finding inner peace has never been such a luxurious experience. This magnificent villa is set within the lavishly-designed Bulgari resort, and has 1,300 m² of awe-inspiring space spread over two levels. The alchemy of sophisticated Italian style combined with the graceful traditions of Bali is magical.

Breathe in the fragrant Jasmine and Frangipani as you relax by the 20 m pool in your cliff-top paradise. Watch the sun rising and setting over the Indian Ocean from the exquisite outdoor living area with its pavilions and lush gardens. And if you want to feel the warm waves lapping your feet, just take the funicular down to the sandy beach.

The Bulgari Villa strikes a harmonious balance, offering private accommodation with superb hotel-style amenities on hand if you need them. Housekeepers come in twice a day, and a butler is at your service. Just beyond the walls of your private sanctuary are an elegant bar, two enticing restaurants, a business centre and a full service spa.

As for the interior of the Bulgari Villa: "stunning" doesn't even begin to describe it. Entering through the double height hallway, you are surrounded by ethereal light and natural materials such as volcanic stone, mahogany and bamboo.

Opulent fabrics and contemporary furnishings characterise the living area, with its private bar, a cinema room, designer kitchen and dining room. Indulge in a deep, relaxing massage in the Bulgari Villa's private spa treatment room, and chill out in the light-filled bathrooms, stocked with Bulgari toiletries, robes and slippers.

宝格丽别墅，高雅的意大利魅力与巴厘岛精神美感的结合。它是巴厘岛悬崖上的璀璨珠宝，是世界萤火虫独特奢华别墅建筑集的一部分。

如此奢华的体验令人内心无法平静。这栋壮丽的别墅位于奢华的宝格丽度假村，占地 1 300 平方米，令人敬畏的空间分布在两个楼层上。精致的意大利风格魅力与巴厘岛优雅的传统风格相结合，十分神奇。

在崖顶乐园的 20 米水池中放松自我，闻着茉莉花和赤素馨花的香味，在被亭子和郁郁葱葱的花园包围的雅致户外生活空间观看印度洋的日出日落，倘若你想让温暖的海浪拍打双脚，可以乘坐缆车前往沙滩。

宝格丽别墅协调匀称，不仅提供私人住宿空间，还有便利高档的酒店设施。房屋女管家每天来两次，还有一位男管家为你服务。优雅的酒吧、两间迷人的餐厅、商业中心和服务周到的水疗馆与房屋只有一墙之隔。

宝格丽别墅的室内设计可谓"惊人"，无法言表。进入两倍高的大厅，你将置身于飘渺的灯光和天然材料中，如火山岩、红木和竹子。

丰富的织物和当代家具定义了起居空间，包括私人酒吧、电影间、专门设计的厨房和餐厅。在宝格丽别墅的私人水疗空间享受按摩，享用梳妆用品，穿上长袍和拖鞋，在光线充足的浴室中放松自我，体验极致的奢华生活。

VILLA AMANZI
阿曼兹别墅

Contributor:
Awesome Villas

Location:
Phuket, Thailand

Villa Amanzi is nestled high up against a lush green hill with each of her floors hugging the landscape and providing commanding views of Kata Noi and the Andaman Sea, while a series of lush hills frame the bay to provide guests with their very own secluded corner of paradise. Rented to only one group at a time to ensure guests' privacy, Villa Amanzi's location, layout and staff services make it an ideal choice for large families or groups of friends.

The villa offers every modern convenience and showcases magnificent sunsets and the white crescent beach below is inspiring. The luxury villa also features an inviting azure infinity swimming pool as well as spacious living, dining and lounging areas decorated with a mélange of Indo-Asian antiques and comfortable furniture.

Comprising 2 master suites, 2 guest bedrooms and a contained guest suite complete with private lounge, flowing space and privacy is evident throughout.

http://m.acs.cn/3u4125/

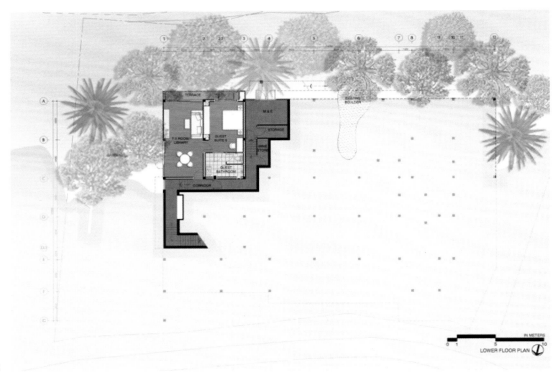

Lower Floor Plan

Main Floor Plan

　　阿曼兹别墅高耸在山脚，背靠葱郁的山头，每层楼都拥有壮丽的景观，包括小卡塔海滩和安达曼海风光。茂盛的小山似乎给海湾镶了边，给客人提供了隐蔽的个人天堂。别墅一次仅租给一组游客，确保了客人的隐私。无论是别墅的位置，还是布局与服务，都是大家庭和朋友度假的理想选择。

　　屋内设有现代生活设备，十分便利，还能欣赏壮观的日落风景，屋下的月牙形白色海滩着实振奋人心。这栋奢华的别墅还设有迷人的蓝色泳池。印度、亚洲混合风格的古董和舒适的家具装饰了宽敞的客厅、餐厅和休息空间。

　　别墅设有两间主套房、两间客房、一间配有私人休息空间的客房套房。每个空间都很流畅，隐私也得到了充分的保证。

Upper Floor Plan

VILLA YIN
"阴"别墅

Contributor:
Awesome Villas

Location:
Phuket, Thailand

Enter Villa Yin and discover the treasures of one of Phuket's true luxury rental gems. This palatial work of art — inspired by respected Thai architect Charupan Wiriyawiwatt of Hong Kong-based Naga Concepts — mixes elements of classic Thai architecture and a dazzling collection of contemporary art with an ultra–luxury design approach.

Pamper your soul in seclusion and privacy at Villa Yin. Situated on Kamala headland's water edge, this spectacular half-acre site overlooks the breathtaking turquoise waters of the Andaman Sea and features a unique private stairwell to a natural, shallow swimming enclave in the sea below.

Delight in the villa's indoor-outdoor tropical living plan which showcases four elegant bedrooms and a 18-meter marble pool that wraps around expansive entertaining platforms along the island's most prestigious west coast location. Villa Yin is an ideal luxury retreat for families and private and corporate groups.

Escape into a world of ultimate beauty. Villa Yin fuses bold antiques such as 14th and 15th century Thai Buddhas, and ancient Burmese gongs, with modern works by artists Xiu Ciu Wen, David LaChapelle and Andy Warhol.

The attention to detail is astounding. Sleek materials such as dark hardwood terracing, marble and granite highlight the unique design statements introduced throughout the property: plate glass vanities and shower screens with pebble and bamboo inclusions, iridescent beaded bathtubs and glass stairways.

Villa Yin's five Thai-style pavilions — the entrance, guest, outdoor, main and master pavilions — are all connected by footpaths, but secluded by design. The estate also features a private stairwell from the cliff-edge pool to a natural, shallow swimming enclave in the sea below.

The state-of-the-art Villa Yin kitchen from Minotti, Italy, winner of Wallpaper Magazine's Kitchen of the year 2006 and other luxury features such as wine cellar, sauna, massage room and home cinema all add to the beautifully appointed outdoor dining and chill out area with a 19-meter marble pool and expansive entertaining platforms.

http://m.acs.cn/3u4119/

"阴"别墅是普吉岛真正的奢华租赁宝地之一，步入别墅，便能发现它的宝藏。如宫殿式的艺术品建筑是泰国德高望重的建筑师Charupan Wiriyawiwatt受到香港那伽理念的启发，用极为奢华的设计手法将传统的泰式建筑元素与各种当代艺术融合在一起。

别墅坐落在卡马拉海岬边缘，其隐逸使人心旷神怡。其场地只有半英亩，却拥有壮观的风景，能欣赏迷人的蓝绿色安达曼海。此外，别墅还拥有一个独特的私人楼梯井，通往低于海平面的天然浅泳池。

别墅的室内与室外热带居住空间设有四间文雅的卧室和一个18米长的大理石水池，水池旁的露台沿岛屿中最有名的西海岸延伸，十分宽阔，可供娱乐，一切布置令人欣喜。"阴"别墅是家庭、个人和团体的理想奢华度假场所。

隐逸于极致的美丽世界之中，别墅引用了鲜艳的古董，如14、15世纪的泰国佛陀、古缅甸锣，还有当代艺术家Xiu Ciu Wen、大卫·拉切贝尔(David LaChapelle)和安迪·沃霍尔(Andy Warhol)的作品。

别墅采用了光滑的材料，如深色的硬木楼梯，大理石和花岗岩则突出了贯穿别墅的独特设计风格：平坦的玻璃板盥洗台、镶有卵石的竹框浴室隔板、彩色串珠浴缸和玻璃楼梯。其细节精致，令人赞叹。

该别墅共由五座泰式亭楼组成。它们分别是入口亭楼、客房亭楼、户外亭楼、主亭楼以及主卧亭楼。它们由小径相连，又从设计上被分隔开来。一处隐蔽的石梯从崖壁泳池通向海平面的天然浅水池。

厨房中的先进厨具是由2006年《壁纸杂志》中厨房部分的奖项得主意大利米洛提(Minotti)提供。其他豪华设施，如酒窖、桑拿室、按摩室和家庭电影室，也都被添设到布置精美的户外用餐和乘凉空间内，当然空间内还设有19米长的大理石泳池和宽阔的娱乐场地。

VILLA NAMASTE
纳玛斯蒂别墅

Contributor:
Awesome Villas

Location:
Phuket, Thailand

Villa Namaste offers one of the best views in Phuket, surveying the wide expanse of the sparkling Andaman sea and the famous dome shaped mountains of Pang Na in the distance — a majestic backdrop to the entertaining pageant of yachts, long tail boats and the glistening swirl of Bang Tao's glorious beaches by day, and the twinkling lights of Bang Tao and Surin by night.

This stunning home is luxuriously finished with furniture and lifestyle decor to the owners' exact specifications; marble floors enhance the elegant aura throughout.

Villa Namaste's spacious interiors, its perfect feng shui, and its coveted locale on a clifftop make this new villa the epitome of Thai elegance and gracious living. In a word: Exceptional.

Please explore this home further and imagine yourself as one of the privileged few who will soon be enraptured by her grace, serenity and vacation perfection.

http://m.acs.cn/3u4126/

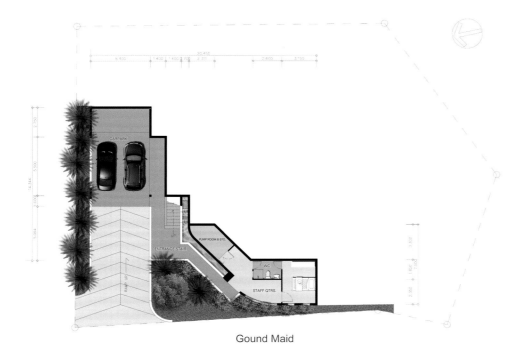

Gound Maid

纳玛斯蒂别墅拥有普吉岛最美丽的风光，远观广阔的安达曼海波光粼粼的海面，以及远处著名的穹顶状庞那山（Pang Na）。这些风景恰好成为乘坐游艇、长尾船游行的壮丽背景，甚至是班涛海滩白天的粼粼波光，班涛和苏林夜间的闪烁光芒也要在它的衬托之下，才能尽显魅力。

别墅内部装饰奢华，令人赞叹。室内陈设和生活饰品无不符合房主要求的具体规格。大理石地面增强了空间的优雅气质。

纳玛斯蒂别墅内部空间宽敞，符合风水习惯。别墅坐落在崖顶，使得别墅成了泰式优雅和高雅生活的缩影。总而言之，别墅独一无二。

深入探索别墅，把自己想象成拥有特权的少数人之一，沉迷于她的优雅、宁静之中，享受这一理想度假场所。

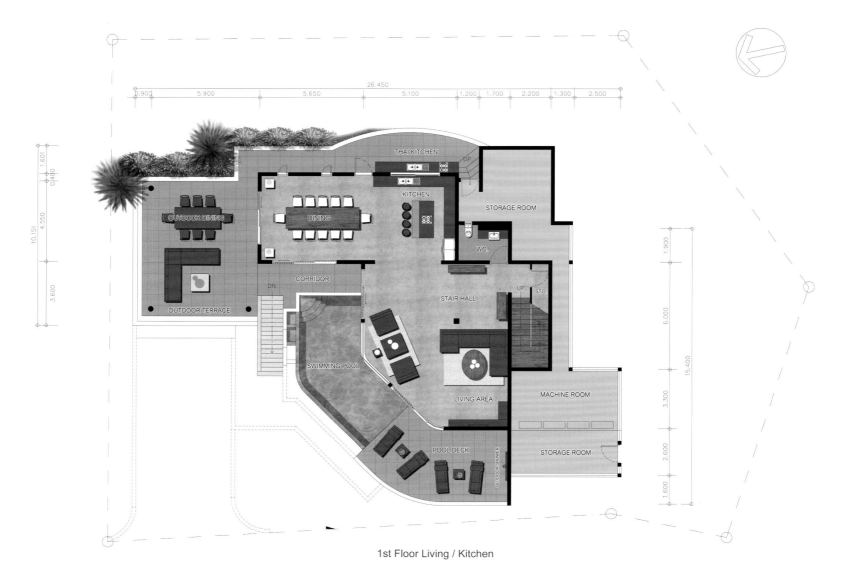

1st Floor Living / Kitchen

2nd Floor Bedroom

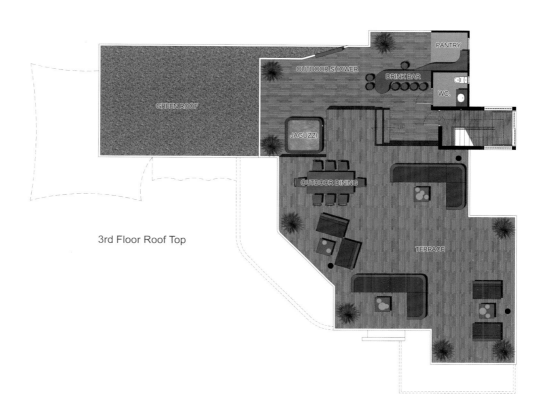

3rd Floor Roof Top

VILLA BENYASIRI
本雅思瑞别墅

Contributor:
Unique Retreats

Location:
Phuket, Thailand

Photography:
Lesley Fisher

Villa Benyasiri is perched atop the highest point of the Samsara headland, affording guests breathtaking vistas across Naka Lay Bay and the glistening Andaman Sea beyond. Its wide sweeping footprint offers an appealing choice of open alfresco spaces and private indoor sanctuaries.

The villa's main entrance opens into the spacious living room, where double-height vaulted ceilings and glass facade panels perfectly frame the postcard-perfect views below. Clear doors fold open onto the infinity pool, which is surrounded by an oversized sun deck and furnished with cushioned loungers for daytime repose. Off in the corner, a narrow path through the surrounding foliage brings guests to a secluded open air "sala" pavilion, a secret and exclusive relaxation space and viewing platform, perfect to retreat to with a good book.

Whites, grays and blacks form an ultra-contemporary color palette at this sophisticated residence, which is also peppered with exquisite Asian artwork and furnishings. Inside on the main level, minimalist décor belies the plush furnishings and inviting ambience, while sleek design lines subtly define the cleverly allocated living spaces. A modern, open-plan show kitchen blends seamlessly with a long, single-piece dining table, complementing the alfresco dining options, the latter including a formal outdoor table and casual lounge area.

Varied bedroom configurations cater to a range of guest needs, especially for multi-generational families or groups of friends sharing the villa. The lower level boasts most of the villa's bedroom sanctuaries, save for one located on the upper entrance level. Tranquil secluded spaces, private terraces and indulgent bathtubs complete the list of private pleasures.

Entertainment rooms are similarly well thought out at Villa Benyasiri, providing inspiring spaces conducive to family gatherings and group activities. A table tennis table near the swimming pool begs a friendly match or two made unique by the tropical backdrop, while for a more relaxed get-together, a private cinema room invites guests to huddle in front of the large projection TV and enjoy the latest blockbuster film.

http://3d.acs.cn/2015/4147/

BENYASIRI - Villa 15
ENTRANCE / POOL LEVEL
Approx. 490 sq.m.

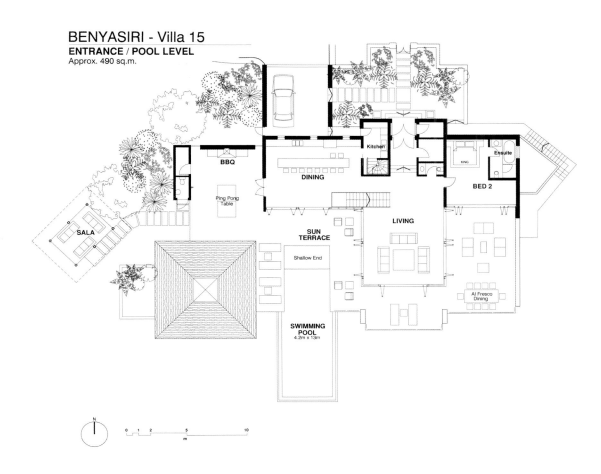

BENYASIRI - Villa 15
GARDEN LEVEL
Approx. 433 sq.m.

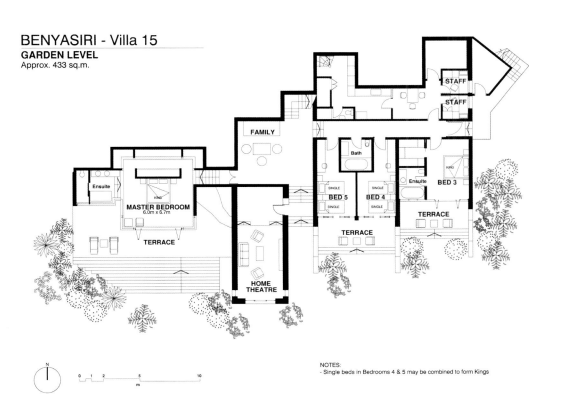

NOTES:
- Single beds in Bedrooms 4 & 5 may be combined to form Kings

本雅思瑞别墅坐落在圣莎拉海岬的最高处，拥有从纳卡雷海湾到安达曼海的迷人风光。开放的设计风格形成了开阔的户外空间和安静的室内私人空间。

踏过别墅的主门，便来到了宽敞的起居空间，两倍高的拱形屋顶和玻璃墙面赋予空间完美的户外风景。透明折叠门直接向无边水池敞开，宽阔的阳光甲板围绕水池而建，上面摆放了软垫躺椅，可供白天休息。别墅一角的小径穿梭在茂密的树林中，把客人引到隐蔽的户外亭子中，这里不仅是秘密又隔绝的放松空间，也是欣赏风景的平台，还是捧书阅读的理想场所。

白、灰、黑是这栋高档居所的主色调，具有超级现代化的风格。精致的亚洲艺术品和家具点缀着室内空间。在主层中，极简主义风格的装饰品与毛绒家具竞相争艳，营造出引人入胜的气氛，光滑的设计线条定义了分配巧妙的起居空间。现代的

开放式厨房与单一的长形餐桌连为一体。此外，户外用餐空间使别墅整体更加完善，其中包括正式的户外餐桌和休闲的休息区域。

各种卧室配置能满足多样的客户要求，适合多代家庭和朋友团体共享。除了设在上面入口层的卧室，底层囊括了别墅的其他大部分卧室。该层拥有安静隔绝的空间、私人露台和大型浴缸，增添了私人乐趣。

同样，设计巧妙的娱乐空间诱惑着人们去参加家庭聚会和团体活动。游泳池旁的乒乓球台让人忍不住想要打一两回合友谊赛，试想一下，在热带风景中打球是如何独特。若想拥有更轻松的聚会，则可聚在私人影院中，大屏电视和最新大片能让人身心放松，尽情享受。

VILLA FAH SAI

法萨伊别墅

Contributor:　　　　Location:　　　　　Photography:
Unique Retreats　　Phuket, Thailand　Lesley Fisher

Villa Fah Sai blends the striking contemporary design elements of Frank Lloyd Wright with the refined comforts of a luxury tropical retreat. Having formerly graced the pages of The New York Times as one of its featured "Great Homes of the World", this is a truly unique and elegant seaside residence.

Making the most of its majestic location, this modern two-level home utilizes strong horizontal lines to provide uninterrupted panoramic views from all its spacious rooms. The grand entrance opens into a wide living room, the space naturally lit with the sun's rays radiating through the paneled glass facade, which can be opened to draw in the balmy ocean breeze. A plush, oversized sofa provides front row seats for the eyes to gaze over an onyx-tiled reflecting pond that seamlessly melts into Naka Lay Bay and the shimmering Andaman Sea beyond.

The clean lines and wide airy spaces of the upper level rooms create unique elements that define the house. Walking through the main atrium, an elegant dining area leads into a bright kitchen, the latter fully equipped with the latest gourmet appliances. Casual bites can be enjoyed in a cozy built-in breakfast nook, while an adjoining terrace provides diners with tempting alfresco options or more formal culinary journeys with a view.

As the day winds down, guests can retire into one of five private bedrooms, each one designed as a personal retreat complete with ensuite bathroom. Apart from the master bedroom, which is nestled into a quiet corner on the upper level, all bedrooms open onto the villa's expansive lower level poolside veranda, the perfect place to congregate for a sunset cocktail or simply between dips in the enticing 18-meter fresh water infinity pool.

Entertainment options include a game room, equipped with flat screen TV and an Xbox 360 for friends or family to enjoy. A treadmill is also stationed in this room, strategically positioned to make the most of the alluring sea views. For the best in leisure luxury, one of the bedrooms on the lower level can also be converted into a home spa with professional therapists on call to administer blissfully, relaxing treatments. And, an indoor Jacuzzi adds another option for an indulgent escape.

Villa Fah Sai is tended by two discreet live-in staff: one chef and one housekeeper. The Samsara Estate front office team and villa concierge complement the villa's permanent staff to organize customized requirements and satisfy any holiday whims.

http://3d.acs.cn/2015/4148/

FAH SAI - Villa 8
POOL LEVEL
Approx. 538 sq.m.

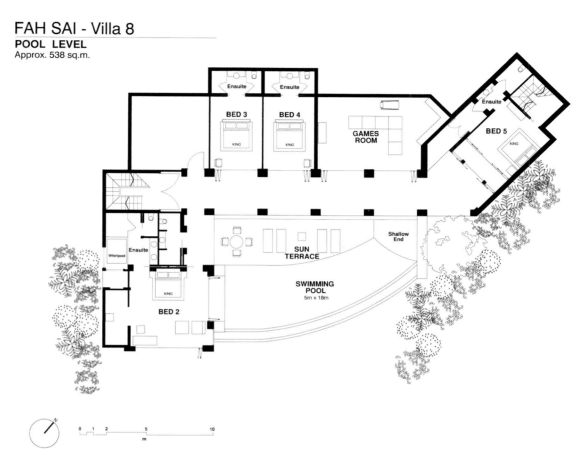

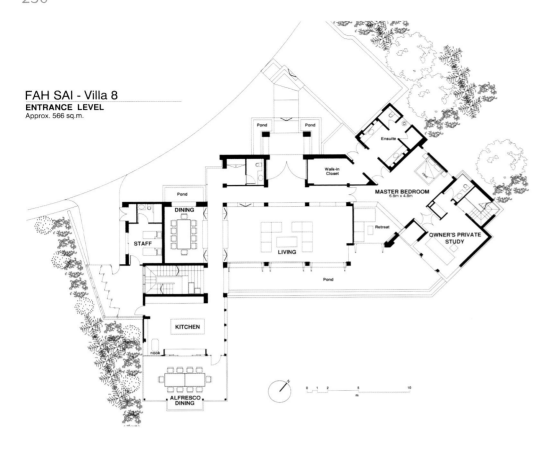

被《纽约时代周刊》专栏评为"世界上壮观的房屋"的法萨伊别墅将弗兰克·劳埃德·赖特当代惊艳的设计元素和奢华热带度假胜地的舒适感融合在一起,成为一栋真正独特、优雅的海边别墅。

这栋现代化双层房屋的设计充分利用它的地理位置,令人可从别墅的每一个宽敞房间欣赏到全景。进入别墅大门,就可以看见宽敞的客厅,这个空间被穿过玻璃墙的太阳光线照亮,还可享受温暖的海风。除此之外,坐在豪华大沙发上可以看见用黑玛瑙平铺的泳池,水池似乎与纳卡雷海湾、耀眼的安达曼海连为一体。

楼上空间线条清晰、宽敞通风,营造出了独特的氛围。穿过中庭,就可以看到高雅的用餐区和明亮的厨房,厨房配备了最新的厨房用具,在惬意的步入式用餐区中可以悠闲地享用食物,紧邻的大阳台给用餐者提供了户外用餐的选择,也可一边欣赏风景,一边享用正式的美食之旅。

别墅设有五间卧室,每间都是私人天堂,并且都带有浴室。当夜幕降临,客人可以在卧室休息。除了二楼安静角落处的主卧外,所有的卧室都面向别墅开阔的泳池边阳台,这正是日落时分品尝鸡尾酒的绝佳场所,当然也可浸泡在18米宽的泳池里,沉醉于清凉的池水中。

别墅还提供了娱乐空间,比如配备了大屏电视和Xbox 360的游戏室,最适合家庭和朋友一起享用。空间内还摆放了跑步机,其摆放位置恰到好处,使你能够欣赏到迷人的海景。为了实现休闲与奢华,同在此层的一间卧室能够转化成家庭水疗室,随叫随到的专业技师将给你提供愉悦轻松的服务。另外,室内的按摩水缸也是放纵享受的又一选择。

法萨伊别墅交由两位入住员工小心管理:一位是厨师,另一位是管家。"圣莎拉"房产的前厅管理人员和别墅管理员是别墅的永久员工,专门处理顾客的个性化需求,满足其任何节日要求。

VILLA HALE MALIA
黑尔·玛丽亚别墅

Contributor:
Unique Retreats

Location:
Phuket, Thailand

Photography:
Lesley Fisher

Villa Hale Malia is a refined holiday retreat that makes full use of its magnificent oceanside location. With a contemporary approach to tropical minimalism, a prolific use of natural materials blurs the lines between in- and outdoor living, while spectacular views create a soothing, natural canvas on which to paint a perfect personalized holiday.

Guests descend down a bold staircase into the main living room, an airy double-height atrium surrounded by clear glass panels that draw in the ocean panoramas and verdant vistas. On one side of the villa's wrap-around terrace, a built-in barbecue holds center-court in the outdoor dining area. On the other side, an expansive sun deck frames the infinity pool and leads to an elevated outdoor pavilion, encouraging guests to pause and breathe in the calm surroundings.

Contemporary furnishings are softened with an earth-toned color palette set against light wood floor panels, with select pieces of Asian decorative woodwork adding an organic touch to the interiors. A fully equipped open-plan kitchen extends into the generous dining area, the entire space made unique by a granite island breakfast counter. This area forms the villa's main indoor communal living area, enclosed in glass doors that fold open onto the generous poolside terrace.

Wellness-oriented facilities provide guests with a range of enticing options to re-center the body and soul. Separate sauna and steam rooms offer moments of serene privacy; alternatively, the relaxation sala can be transformed into a yoga pavilion, perched majestically overlooking the ocean. Further additions such as a lower level meditation corner and oceanfront lawn provide a zen-like atmosphere to gently focus the mind.

http://3d.acs.cn/2015/4151/

HALE MALIA - Villa 5
POOL LEVEL
Approx. 560 sq.m.

HALE MALIA - Villa 5
GARDEN LEVEL
Approx. 500 sq.m.

NOTES:
- Single beds in Bedroom 3 may be combined to form a King

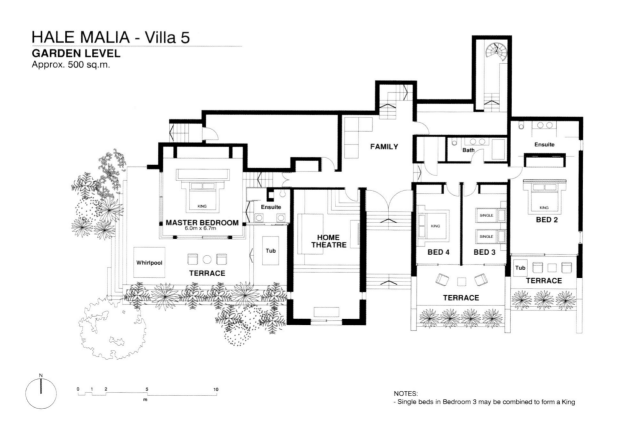

黑尔·玛丽亚别墅位于海岸边，优雅的度假寓所充分发挥了其壮观的海滨特色。当代的手法与热带极简派艺术融合，并充分使用天然材料，模糊了室内与室外生活空间。同时，壮观的景色形成了舒心的天然画布，预示着完美的个性化假期。

从醒目的楼梯走到主客厅，开敞的两倍高中庭被透明玻璃幕墙包围，把海洋全景和狭长的翠绿风景引入室内。在别墅一侧的环绕式露台，内置的烧烤空间成为户外用餐空间的核心。另一侧是开阔的日光甲板，通往被抬高的户外凉亭，游客在此能欣赏无限延伸的水池，静谧的风景不禁让人停住脚步，沉醉于其中。

现代家具采用了大地色调，与浅色木地板形成对比，显得十分柔和，精心挑选的亚洲木饰品给室内空间增添了有机的一笔。装备齐全的开放式厨房延伸至开阔的用餐空间，独立式花岗岩餐台使整个空间变得独特。这里便是别墅室内的主要公共生活区域，空间被折叠玻璃门包围，面向开阔的池边露台。

健康型器械给客人提供了多项诱人的选择，帮助你重塑身心。独立的桑拿和蒸汽浴室给人以片刻的平静和隐私。或者把休息大厅转变成瑜伽亭，亭子庄严地屹立在高处，俯瞰海洋风景。另外还有楼下的冥想区、海滨草坪，禅式氛围让人慢慢地集中精神。

Intelligently thought out space allocations create versatile options for guests. Ensconced in tranquility, four seaview bedrooms in varying configurations come with added personal luxuries including private terraces and outdoor bathing options. Families or groups of friends can enjoy quality time together with a movie night in the TV room, by gathering on the pool deck for a sunset cocktails or comparing their favorite holiday reads in the plush study.

巧妙的空间分配给客人提供了多样的选择。四间卧室面向海洋，十分安静。室内设施丰富，增添了个人奢侈品，比如私人露台和户外沐浴设施。家人或朋友能一起在电视房中享受电影之夜和高品质时光，日落时分欢聚在水池露台，品味鸡尾酒，分享自己最喜欢的假期，尽享奢华。

BAN MEKKALA
班麦卡拉别墅

Contributor:
Zekkei Collection

Location:
Koh Samui, Thailand

Ban Mekkala has been designed to accentuate comfortable and convenient living in a tropical paradise. From the exterior, the imposing walls lend little to the fact that this is truly a contemporary house that has combined its location and spacious design with the luxuries of modern technology and quality furnishings to create a hideaway for those seeking a relaxing retreat, or fun getaway.

Upon entering the villa compound, you are enticed into the residence over a peaceful stepped pond filled with Koi fish and the stunning view through the property to the glistening ocean in the distance. The luxury residence is furnished with modern Thai décor, designed to accentuate comfortable and convenient living in a tropical environment. An imposing entrance opens to reveal a truly contemporary home.

The main entrance to the villa leads directly into the cavernous interior of the combined living, dining and kitchen areas. This grand room on the second level enjoys stunning views of the natural surroundings from all angles through full length windows running across the expanse of the building. There is an enormous gourmet kitchen, a twelve-seat dining table, a lavish living/entertainment area equipped with a state-of-the-art AV system and plasma TV. There is surround sound technology linked to every part of the house (inside and out), so your favorite tunes can be enjoyed from wherever you are!

http://m.acs.cn/3u4137/

BAN MEKKALA

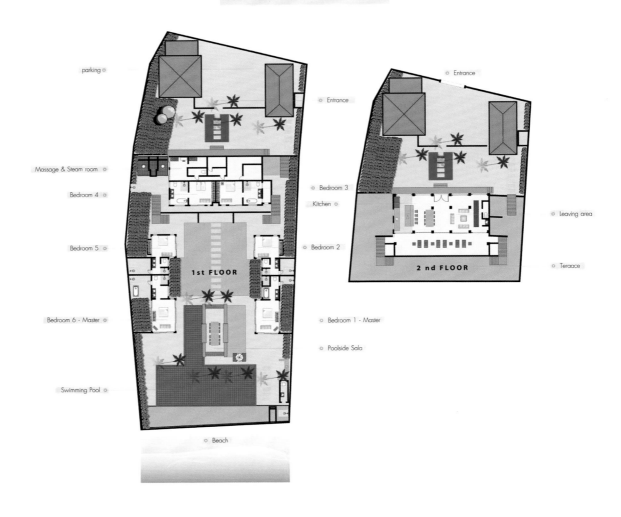

班麦卡拉别墅是一个热带天堂，舒适且便利。从外面看去，雄伟的墙让人很难相信这实际上是一栋因地制宜的现代房子。宽敞的设计、现代化的科技以及高质量家具给想寻求放松、娱乐的人们提供了理想的隐居场所。

别墅前面是平静的水池，池中的步石汀步让人不由自主地踩踏上去，锦鲤徜徉在水池中，而远处则是波光粼粼的大海。豪华别墅内采用了现代泰式装饰物，突出了热带环境中舒适、便捷的生活。走过威武的大门，便是真正的当代家庭。

别墅主入口直接通往幽深的生活、用餐和厨房混合空间。建筑大面积采用了全景玻璃窗，使得在二楼大房间的任何角度都能欣赏到迷人的自然风景。别墅内设施齐全，包括无数美食炊具和12座餐桌。宽敞的休息娱乐空间设有最先进的AV设施和等离子电视。环绕立体声音响设备遍布房子的每个部分（室内和室外），从任何角落你都能欣赏自己喜爱的音乐。

One will reach the ground floor via a staircase that overlooks a waterfall feature flowing into an internal courtyard. This level houses all six bedrooms consisting in three distinctive styles, each coming with a generous en-suite bathroom, and a wide flat panel TV connected to DVD player. These individual bedroom pavilions are built amid a beautifully landscaped garden. The two rooms on either side of the central lawn benefit from an additional outdoor shower in their own private courtyards.

At the far end of the property you'll find the villa's large swimming pool with incredible sunken sun benches. here's an outdoor dining pavilion as well as a barbecue and kitchen for the ultimate outdoor dining experience.

The villa sits on the southern coast of Koh Samui, some 35 minutes from the Airport and 15 minutes from the town of Lamai. In the Summer low tide season when the reef is dry for up to 100m, this opens the beach and reef up to all sorts of other activities such as beach volleyball, football or simply exploring the rock pools and collecting shellfish.

通往一楼的楼梯设置巧妙，可以欣赏瀑布落入内院。该楼层上的六间卧室具有三种显著的设计风格，每间房内都配有浴室、宽屏电视和 DVD 机。私人卧室阁楼设在美丽的景观花园中。在草坪两端的卧室的私人庭院中都带有户外浴室。

别墅的尾部是大泳池，凹式太阳长椅令人不可思议。此处的户外用餐凉亭和户外烧烤、厨房空间将给人非凡的户外用餐体验。

别墅坐落在苏梅岛的南海岸，离机场只有 35 分钟的距离，离拉迈镇也只有 15 分钟的距离。在低潮的夏季，暗礁退水达 100 米时，海滩和礁石给各种活动提供了充足的空间，例如沙滩排球、足球，或是简单的岩池探索和贝壳拾捡活动。

VILLA YANG SOM
杨孙别墅

Contributor:
Awesome Villas

Location:
Phuket, Thailand

Villa Yang Som is nestled high up against a lush green hill with each of her 6 floors hugging the landscape and providing commanding views of Phuket Island and the Andaman Sea, while a series of lush rolling hills in the distance frame Phuket island's most desired position to provide Villa Yang Som's guests with their very own secluded corner of paradise.

Designed to feel like a home away from home, this magnificent villa is decorated with handmade silks & teak furnishings, and is surrounded by lush palms and evident privacy. The villa's handsome accommodations sleep up to 10 guests. The outdoor areas overlooking Surin and Bang Tao beach area are idyllic for watching magical Thai sunsets with cocktails and pre-dinner canapés prepared by the Villa's private chef. A state of the art sound system plus comfortable sofas with over-sized cushions below the towering Thai salas make Villa Yang Som the ultimate chill-out spot by the sea…

As you arrive on street level you are immediately met with an astounding structure that rises triumphantly skywards, towering above you with architectural magnificence.

As you step into the panoramic elevator you are gently ported to the main reception floor where you are greeted with sweeping spaces, water features and an abundance of teak.

This floor also boasts a large games room which can double up as a fifth bed room with en-suite bath room suitable for a large family with school-going children. A Thai style kitchen and a laundry room are also available on this level.

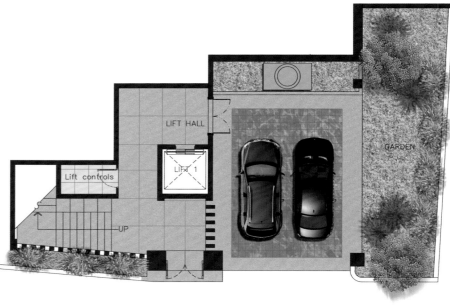

1st Floor Plan

杨孙别墅高耸在葱郁的山上，六层楼都依山而靠，俯瞰普吉岛，远观安达曼海。远处连绵不断的山就像是给普吉岛上令人梦寐以求的土地装上了画框，给别墅的客人提供了隐蔽的天堂一角。

别墅设计旨在给人一种家外之家的感觉，屋内装饰着手织丝绸和柚木家具，而屋外是茂密的棕榈树，显然赋予了别墅隐私。别墅能入住10人以上。户外空间能远眺苏林和邦涛海滩，坐在两片田园式的沙滩上，吃着私人厨师准备的餐前小吃，品着鸡尾酒，欣赏着神奇的泰国日落，最美不过如此。高耸的泰式大厅下，摆放着舒适的超大沙发和软垫，配上先进的环绕立体声音响系统，使杨孙别墅成为海边最好的乘凉地点。

初入此地，建筑奇观立即出现在眼前，高耸的建筑直接冲向天空，十分宏伟。

步入观光电梯，缓缓的电梯把你带入主接待层，而开阔的空间、水景和丰富的柚木结构却涌现在你眼前。

此层还设有大型游戏室，与房间连在一起成为第五间卧室，内部设有配套的浴室，适合有孩子上学的家庭居住。此外还有泰式风格的厨房和洗衣房。

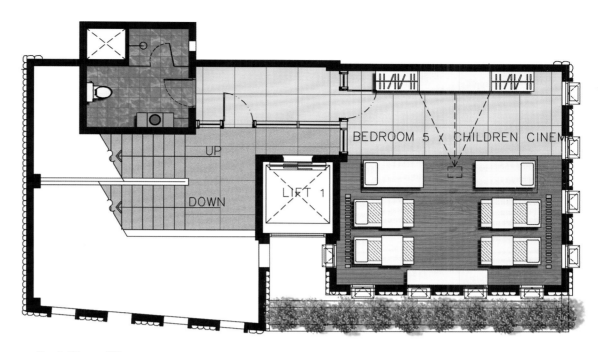

2nd Floor Plan

3rd Floor Plan

The elevator further opens onto an open reception area that comfortably seats up to forty guests. With a long bar at one end it is perfect for private parties and events, where guests can stroll off to the sala at the front and quietly enjoy the panorama of the Andaman Sea as it caresses the shores of Phuket island.

A raised gallery above the reception area leads off from the centre to the 2nd elevator and Master Suite 4; and on the other side, a grand staircase meanders up the main living floor with an 18 meter infinity pool and a Jacuzzi.

随着电梯上升，接下来便来到了开放式的接待区，此空间能舒服地坐下40位客人。一端的长吧台最适合举行私人派对和活动，客人可以从这儿散步至前端的大厅，安静地享受环绕普吉岛海岸的安达曼海的全部风光。

走过接待区上方的抬高式走廊，便从中央区域来到了第二电梯和第四主套房；另一侧的大楼梯蜿蜒上升，穿到主生活层，该层设有18米长的水池和按摩浴缸。

4th Floor Plan

This spacious Main entertainment floor incorporates open plan living, dining and entertainment areas, beautiful stone terraces, an infinity pool, with trance-inducing views and an alfresco dining Sala for sanctuary from the midday sun. Cooled by tropical breezes and permeated by the rhythmic sounds of Phuket, Villa Yang Som provides a sensually engaging environment in which to unwind and rediscover your senses...

At the centre, an intimate sala is raised just above the infinity pool to offer you the best vantage point of the sea and landscape tantalizing your vision. The Sun decks spread out on both sides of the sala and ensure you are never too far from either basking in the warm sun or taking a brief respite from the heat.

宽敞的主娱乐层包括开放式的客厅、餐厅和娱乐区，美丽的铺石露台和无边泳池给人扑朔迷离的景观，户外用餐大厅拥有遮盖，防止中午日光直射。热带轻风给杨孙别墅带来了清凉的慰藉，普吉岛上传来有节奏的音乐，这样迷人的环境，能让人放松，焕发活力……

中央是气氛亲切的大厅，大厅高于无边泳池，给你最佳的观海视角，各种景观也吸引着你的视线。阳光甲板在大厅的两侧，空间充足，无需走远便能在温暖的阳光中沐浴、放松。

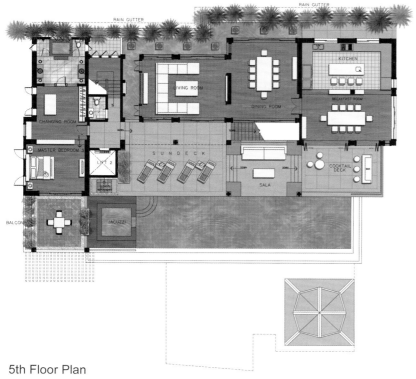

5th Floor Plan

6th Floor Plan

Behind the deck is the living room and dining room, next to which a large and fully equipped kitchen that opens onto a large informal breakfast lounge. From there you can step out to the deck from other side of the sala — complete with a bar and professional service station.

This deck can seat 10 people comfortably — an ideal spot for a quiet breakfast in the early morning hours, or a barbecue dinner during sunset. At the other end of this floor is Master suite 3, complete with an adjoining study and its own balcony, the swimming pool just a mere footstep away.

From the 6th floor the half-moon sala looks over the infinity pool below, and is serviced by an adjoining bar. This floor is accessible by lift from the gallery on the 4th floor. Needless to say, the view is spectacular!

甲板后方是客厅和餐厅，设备齐全的大厨房也就在旁边，敞向休闲式大型早餐室。踏出此处便到了大厅另一侧的甲板。甲板空间设有吧台和专业服务站，使这一空间更加完美。

甲板能轻松坐下 10 人，适合于清晨安静地享用早餐，或日落时的烧烤晚宴。

该层的另一端是第三主套房，旁边是书房和阳台，几步之外就是泳池。

四楼走廊的电梯直接通往六楼，于半月形大厅可俯瞰无边泳池，临近的吧间为你提供服务。风景之壮观，无须言表。

The main master suite is just next to the half-moon sala, complete with a spacious study and panoramic views; a private deck to one side with a Jacuzzi and the double sliding doors open onto a panoramic deck for an overwhelming sense of space and tranquility and unending views.

An adjoining massage sala on this floor is the ideal place with which to indulge in traditional massage therapies that Thailand is world famous for.

On the other side of this floor is the 2nd Master Suite.

Relaxation and personalized service are hallmarks of the Villa Yang Som Experience and guests will find everything they need for a fun-filled stay in this secluded luxury hideaway…

主套房就设在半月形大厅旁边，配有宽敞的书房和全景式的风光。一侧的私人甲板上摆放着按摩浴缸，双滑动门外便是露台。全景式的露台给人以宁静、难以抗拒的空间感和漫无边际的景色。

临近的按摩室给客人提供了理想的按摩空间，传统的按摩疗养令人沉醉，这也是泰国享誉世界之处。

该层另一侧是第二主套房。

消遣和私人服务是杨孙别墅体验中的特色，客人将在这隐蔽的豪华居所中找到自己所需要的一切，让此行充满乐趣，满载而归。

VILLA VIMAN
维曼别墅

Contributor:
Unique Retreats

Location:
Phuket, Thailand

Photography:
Lesley Fisher

Villa Viman is a sophisticated, generous family-style holiday home with a multi-level design that makes full use of the property's dramatic hillside and oceanfront location. This stunning, contemporary villa offers a rich blend of bold colors, striking furnishings and incredible island views.

A centerpiece entrance and staircase descends from the foyer into the main lounge, the Andaman Sea glittering through the vaulted glass facade. Clear doors slide all the way open onto a wide pool terrace wrapped around an infinity pool with sunken whirlpool and furnished with cushioned loungers. A separate, covered outdoor relaxation sala and a dedicated alfresco dining space encourage guests to literally "live the view" during their stay.

Inside the villa, wooden floors, doors and ceilings match the rich tropical surroundings, while Asian inspired artwork, plush furnishings and inspired decorative touches add a sense of individuality to the villa's homey ambience.

Clever design touches such as varied bedroom configurations and generous space allocations add versatile options for larger families and groups. The bedrooms are designed in six individual sanctuaries, each featuring en-suite bathrooms and individual terraces to breathe in the ocean panorama whenever the impulse beckons. Unique additions such as hidden outdoor bathtubs add options for indulgent and exclusive private pleasure.

Younger members of the family are well catered for with a playroom attached to the kids' twin bedroom, where a football table and bean bags provide an entertaining space for children to gather. An equally creative use of the expansive living spaces provides guests at Villa Viman with endless options for personalized diversion, whether it's relaxing in front of a movie on the large projector screen in the AV room, challenging family members or colleagues to a game of table tennis in the games room, or working out on the elliptical trainer while you drift into the coastal vistas.

http://3d.acs.cn/2015/4146/

VIMAN - Villa 2
BASE LEVEL
Approx. 133 sq.m.

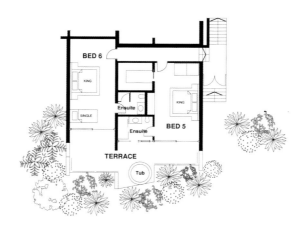

维曼别墅，高端大气的家庭式度假家园，具有多层次的设计，充分发挥了其依山傍海的显著优势。这一令人赞叹的现代别墅采用了大胆、丰富的色彩，混合引人注目的家具，充分利用该岛屿的迷人风光。

从门厅的中央入口进入，穿过楼梯，直接进入主客厅，拱形的玻璃墙外是艳丽夺目的安达曼海。滑开透明玻璃门，宽阔的水池露台围绕着无限延伸的水池，水池内沉落的涡流，别具风采，露台上还摆放了软垫躺椅。独立的带顶户外凉亭，以及专用的户外用餐空间，让顾客在逗留期间真实地体验这番风景。

别墅内，木地板、门和天花板都融入到丰富的热带环境中，亚洲灵感艺术作品、豪华的家具以及点睛的装饰物，给自在的氛围添加了一些个性。

别墅设计巧妙，如各种各样的卧室配置、阔气的空间分配以及提供给家庭和团体的多样选择。卧室分成六个独立的空间，都带有配套的浴室和独立露台，无边无际的海洋无时不刻在冲击着你，召唤着你。别致的附加设施，如隐藏的户外浴缸丰富了客人的私人乐趣。

游戏室设在儿童卧室之间，迎合了家庭中年轻人的需求。游戏室中的桌上足球和豆袋坐垫给孩子聚会提供了娱乐空间。宽敞的生活空间的设计同样富有创意，给维曼别墅中的客人提供了无限的选择，不管是在 AV 室中坐在大投影屏幕前欣赏电影，放松自己，还是在游戏室中与家人或同事对战乒乓球，抑或是沉浸在沿岸风景中，做"空中漫步"锻炼，能满足客人多样的个人需求。

VIMAN - Villa 2
GARDEN LEVEL
Approx. 520 sq.m.

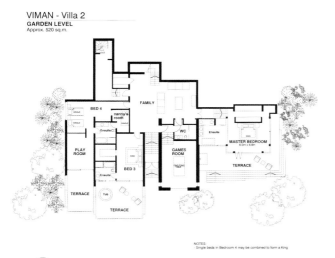

NOTES:
- Single beds in Bedroom 4 may be combined to form a King.

VIMAN - Villa 2
POOL LEVEL
Approx. 628 sq.m.

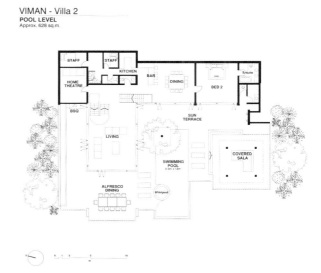

VILLA ROM TRAI
罗泰别墅

Contributor: Unique Retreats
Location: Phuket, Thailand
Photography: Lesley Fisher

Villa Rom Trai is an Asian influenced contemporary holiday home with a striking design that cascades down a gentle hillside to the water's edge, carving out multiple levels of comfortable living spaces and intimate gathering areas, ideal for families and friends. The villa's warm, understated interior decor takes a backseat to the generous views across Naka Lay Bay and out over the Andaman Sea.

The entry staircase descends from the foyer into one corner of the lounge, which opens out and connects to the main communal living spaces. Alternatively, guests can take an internal lift, which connects three of the main levels. Panels in the glass facade slide open onto the wrap-around pool deck, which is furnished with cushioned loungers and alfresco dining coves.

Back inside the villa, a spacious dining area sits under an airy double-height atrium, with a breakfast counter for casual snacks and an enclosed fully fitted kitchen hidden off to one side. Colorful abstract paintings adorn the walls, while dark wood floors and richly toned furnishings bring a sense of sophisticated tropical luxury to the rooms. Tucked away at the far end of the villa, a welcoming, carpeted library room offers a quiet, relaxed space to catch up on some holiday reading.

Four generously appointed bedrooms all come furnished with king-sized beds. Two of the larger bedrooms feature en-suite bathrooms and sprawling private terraces, an ideal spot from which to ponder nature's wonders. The remaining two bedrooms are connecting, making the villa an ideal choice for parents that want to keep an eye and ear out for the younger members of the group at night.

An additional two levels of communal space provide guests at Villa Rom Trai with yet more places to mingle and enjoy the private facilities. The villa's expansive, lawned terrace is ideal for sunset family gatherings or an outdoor game with the kids; while the second swimming pool measures 20 meters and is therefore well suited to swimming laps. Nearby, a steam room and separate sauna offer guests the chance to relax, cleanse and refocus after a day of leisurely pursuits.

Villa Rom Trai is tended by two discreet live-in staff: one chef and one housekeeper. The Samsara Estate front office team and villa concierge complement the villa's permanent staff to organize customized requirements and satisfy any holiday whims.

http://3d.acs.cn/2015/4150/

ROM TRAI - Villa 9
POOL LEVEL
Approx. 555 sq.m.

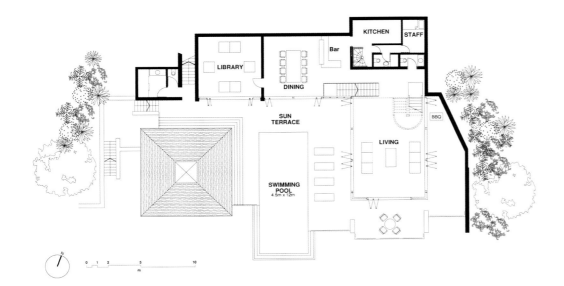

罗泰别墅坐落在水边的平缓山坡上，是受亚洲风格影响的当代假日住宅，其设计引人注目，拥有不同层次的舒适生活空间和亲密的聚会空间，是家庭和三五朋友的理想度假场所。内部装饰温馨朴素，坐享从纳卡海湾到安达曼海的广阔风光。

入口楼梯从门厅延伸至休息厅的一角，敞开后连接了主公共生活空间，顾客也可乘坐连接三层楼的内置电梯。滑动式玻璃墙外是环绕式的水池露台，并摆放了软垫躺椅和户外餐台。

宽敞的餐厅设在两倍高的通风中庭下，一侧是休闲小吃早餐餐台和设备齐全的封闭式厨房。彩色抽象画装饰着墙面，以及黑色木地板和丰富的暖色调家具，给空间增添了热带奢华感。书房隐藏在别墅的深处，屋内铺满地毯，给假日阅读提供了安静、轻松的空间。

四间宽敞的卧室内都设有大号床。其中较大的两间还有配套浴室和不规则私人露台，这是欣赏自然奇观的理想场所。其余的两间卧室是相连的，适合夜间需要照顾小孩的父母。

ROM TRAI - Villa 9
GARDEN LEVEL
Approx. 440 sq.m.

罗泰别墅内额外的两层公共空间给顾客提供了更多空间来参与享受私人设施。宽敞的草坪露台是傍晚家庭聚会或亲子户外嬉戏的最佳场所。第二泳池长20米，同样适用于泳圈游泳。附近是蒸汽浴室和独立桑拿房，让客人在一天的休闲后，放松、净化和调整自己。

罗泰别墅由两位入住的员工精心管理：一位是厨师，另一位是管家。"圣莎拉"房产的前厅管理人员和别墅管理员是别墅的永久员工，专门向客人提供个性化服务，满足其任何度假需求。

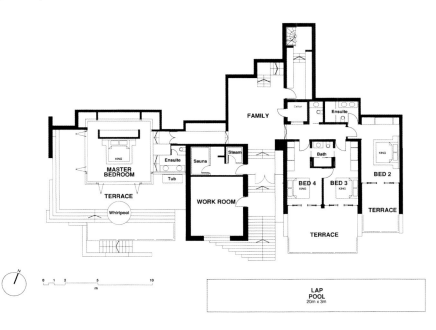

MALIMBU CLIFF VILLA
玛里姆布悬崖别墅

Contributor:
Marketing Villas Ltd.

Location:
Malimbu Bay, West Lombok, Indonesia

Area:
1,000 m²

Set high above the sea on a grassy hillside, this stylish four-bedroom villa has access to its own private rocky beach and huge views that extend across the ocean all the way to the spectacular peak of Bali's mystical Mount Agung.

Fresh and modern in its design, four-bedroom Malimbu Cliff Villa features crisp clean lines with white and grey exteriors to create a light and airy ambiance, while vibrant artwork and colourful fabrics contrast nicely against the neutral furnishings evoking a unique sense of style.

Built with relaxation and comfort in mind, the villa's rooms and living spaces will make guests feel very much at home, however the inclusion of a few decadent features add an exclusive taste of luxury. With an 18-metre infinity pool that appears to drop straight into the ocean; a terrace from which to-die-for sunsets unfold each evening; a master bedroom with its own pool balcony just made for sipping cocktails, and luxurious bathtubs offering some of the best views in the house, it's hard not to get into the laidback Lombok groove at Malimbu Cliff Villa.

While the villa offers peace and privacy in an idyllic, away-from-it-all coastal location, it's only a ten minute drive to Senggigi's bars and restaurants (the villa rate includes a car with driver) and a short boat ride to the gorgeous Gili Islands, famous for their white sands, crystal clear waters and excellent dive and snorkeling sites.

Of course a villa stay is not complete without warm and friendly staff to look after guests' every need, and the team of Malimbu Cliff Villa certainly doesn't disappoint. Fully trained, and with years of experience, the dedicated villa manager, skilled chef, attentive butler and efficient housekeepers will ensure your holiday is truly memorable — from welcome drink to farewell dinner — leaving you longing for more.

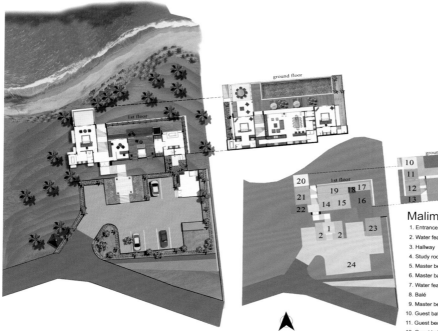

Malimbu Cliff Villa Key Plan
1. Entrance
2. Water features
3. Hallway
4. Study room
5. Master bedroom
6. Master bathroom
7. Water feature
8. Balé
9. Master bedroom terrace
10. Guest bathroom
11. Guest bedroom
12. Guest balcony
13. Downstairs balcony
14. Media/Entertainment room
15. Downstairs guest bedroom
16. Downstairs guest bathroom
17. Swimming pool
18. Sundeck
19. Living & dining room
20. Guest lavatory
21. Service area
22. Kitchen
23. Staff area
24. Parking area

时尚的玛里姆布悬崖别墅高耸在绿草茵茵的山坡上，内部设有四间卧室，拥有私人岩滩和开阔的风景，从海洋延伸到巴厘岛上神秘壮观的阿贡山山峰，美不胜收。

玛里姆布悬崖别墅有着清新现代的设计风格，清晰干净的线条与灰、白的外壁营造出明亮、通风的环境。生机勃勃的艺术品和色彩缤纷的织物与中性的室内陈设形成巧妙的对比，形成一种独特的风格。

用放松和舒适为宗旨打造的别墅将给住客家的感觉，不管是房间还是起居空间，几处休闲空间增添了独特的奢华感。18米长的无边泳池，似乎直接流向海洋；日落每晚在露台上展示风采；主卧室带有私人水池阳台，在此享用鸡尾酒最适合不过了。豪华浴缸形成了屋内最美的一部分风景。在龙目岛的玛里姆布悬崖别墅中不难放松身心。

别墅拥有田园般的宁静和隐私空间，十分钟的车程就能到达圣吉吉酒吧和餐厅（别墅提供车和司机）；乘坐船只，一会儿就能到达美丽的吉利岛，可欣赏白色的沙滩、清澈的海水，享受绝佳的潜水和跳水活动。

当然，别墅还拥有热心、友好的工作人员，照顾每位客人的需求。训练有素、经验丰富的玛里姆布悬崖别墅的团队不会令客户失望。兢兢业业的别墅经理、专业的厨师、专注和高效的管家将给你一个值得回味的假期，从迎接酒会到送别宴，无一不让你回味终生。

VILLA LUWIH
露维别墅

Contributor:
Marketing Villas Ltd.

Location:
Pererenan village, Canggu, Bali

Area:
600 m² (Villa); 1,000 m² (Land)

Villa LuWih has been thoughtfully designed to capitalize on its prime location overlooking Pererenan Beach on Bali's dramatic and unspoiled south-west coastline. Angled towards the ocean and the spectacular sunsets, three floors of stylish and luxurious accommodation are fronted by acres of glass to ensure that wherever you are in the villa, you won't forget where you are in the world.

This congenial villa offers a dazzling choice of adaptable living space. From the slick poolside bar to the state-of-the art media room with its giant 60-inch TV, Playstation and surround-sound, guests of all ages and all interests will find a place to please them. This flexibility and abundance of indoor and outdoor living space makes Villa LuWih an ideal choice for families who want to stay and play together or friends looking for the best of both worlds — peaceful spots to spend their day and social spaces to party the night away.

The six bright, spacious bedrooms, with a snug on the ground floor offering a seventh bedroom option, are spread across all three floors, ensuring plenty of privacy and the convenience of being able to group families together in separate parts of the villa. Bathrooms, all ensuite, are also big and bright, and fitted with large tubs, standing showers and lots of windows to let the views and sunlight flood in. The villa is fully air-conditioned, but most of the glass-fronted rooms can be opened widely to the ocean breezes if tropical outdoor living is your thing.

As part of its super-flexible layout, Villa LuWih offers guests a choice of swimming pool. At ground level is a splendid 16-metre lap pool, with a shallow end for the kids, a shady relaxation balé, and several pairs of brightly cushioned loungers alongside. One level up, flanked by the snug media room and an indoor/outdoor sitting area, is the second pool — smaller than the first, but with the best views in the house from its glass fronted depths.

A team of graceful and welcoming staff will take care of your every need at Villa LuWih. Your villa manager will happily help with tour bookings and restaurant reservations, although with a skilled private chef at your disposal, you may be perfectly content just to stay "home" and enjoy delicious meals selected from the villa's suggestion menu.

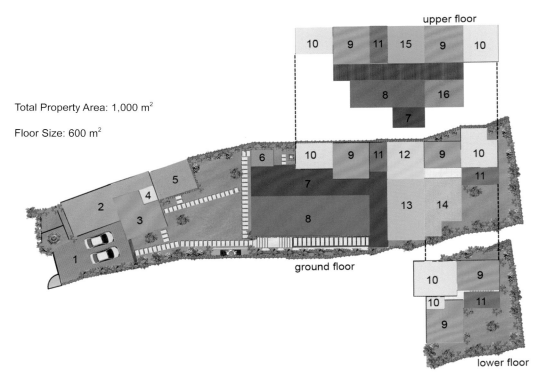

Total Property Area: 1,000 m²

Floor Size: 600 m²

1. Entrance / Carport
2. Staff Office
3. Guest Pavilion Bedroom
4. Guest Pavilion Bathroom
5. Bar
6. Bale
7. Pool Deck
8. Swimming Pool
9. Bedroom
10. Bedroom
11. Stairs
12. Kitchen
13. Living Area
14. Dining Area
15. Media Room
16. Upper Deck

露维别墅设计精妙，发挥了地理位置的优越性，俯瞰普瑞勒南海滩风光，此海滩位于巴厘岛西南海岸，风景保存良好，引人入胜。别墅依一定角度而建，面向海洋，能够欣赏壮观的日落景观。别墅分为三层，都采用了奢华的现代风格，前端是大面积的玻璃墙，保证你不管身处别墅何处，都能感觉到自身处在世界之中。

合乎人意的别墅拥有众多的生活空间可供灵活选择。从池边光滑整洁的吧台，到装配先进的媒体室无一不不使人振奋。媒体间内还设有 60 英寸大屏电视、游戏机和环绕声音响设备，能满足不同年龄、不同爱好的群体的需求。宽敞、灵活的室内外居住空间，使露维别墅成为家庭居住和嬉戏的理想场所，这里白天可充当安静的度假空间，夜间可充当热闹的社交空间，对同时具备这两项需求的朋友同样适合。

别墅拥有六间宽敞、明亮的卧室，一楼的雅间可以作为第七间卧室。卧室分别设在三个楼层，保证了其私密性和便利性。而别墅的间隔区也可作为家庭成员聚会的地点。卧室都带有宽敞明亮的独立浴室，内部设有大浴缸、淋浴设施，多个窗户使得室内空间阳光充足。别墅内装有空调系统，但装有玻璃窗的房间都朝向大海，如果偏爱热带户外气息，便可以让海洋微风吹入室内。

别墅超级灵活的布局中包括了一层的泳池，长达 16 米，一端还专为孩童打造了浅水区，池边还设有带顶的放松区，以及几对亮丽的软垫躺椅。更高一层设有媒体间、室内和室外休息区。此处还有第二泳池，虽然其比第一泳池小，但拥有更美丽的景色，透过玻璃，风光无限。

别墅拥有热情优雅的服务人员，将照顾好你的每一个需求。别墅经理也会热情地帮你规划旅程和餐厅。熟练的私人厨师可随叫随到为你服务，你还能从推荐菜单中任意挑选美食，别墅将给你宾至如归的感觉。

BAAN HINTA
巴安欣塔别墅

Contributor:
Zekkei Collection

Baan Hinta sits on a unique site cascading out over a fine sandy beach. The iconic roof structure provides dramatic lofty spaces of up to 10 meters over the main living areas. From indoors to outdoors, each space features luxurious proportions and expansive sea views.

Noted for breath-taking architecture, this five-bedroom villa features a spacious pool, lounge sala and a unique outdoor cinema. The dominant architectural theme is an iconic roof structure — clad in cedar shingles, these swooping roofs provide dramatic lofty spaces of up to 10 meters over the main living areas, reducing to more intimate scale over the bedrooms. Natural rock structure in the bathroom and outdoor showers add amazing natural touches to the overall earthy interiors. Another amazing highlight is the generous glass wall in the kids bedroom — it sees through to the swimming pool to simulate a private aquarium!

Entering the villa one would be greeted with the ultra-spacious, split-level, wood-clan living and dining area that opens to stunning tropical views. Each space flows seamlessly into another, from indoors to outdoors, with luxurious proportions and expansive views of the beach and ocean. Relaxation space abound, from the lounge sala furnished with two large couches and complete home entertainment system, landscaped poolside terrace with HD outdoor cinema to the spacious dining room that seats 20. The open-plan western kitchen comes with a full host of LG and Siemens appliances for professional cooking to do the huge wooden dining table justice.

There are five luxuriously furnished double bedrooms with king-sized beds, all of which equipped with comprehensive home entertainment and enjoy enviable sea views. The bathrooms feature free standing bathtubs and separate shower cubicles, plus his and her washbasins. Inside the fitted wardrobe you would find plush bathrobes and slippers for world-class comfort. One of the master bedrooms has direct access to the beach. The bedrooms on 2/f are connected by a communal terrace that features an additional outdoor bathroom.

The living area opens directly onto the sun deck, furnished with soft daybeds, and the generous private pool. Right next to the pool, a projector and screen create a unique outdoor cinema experience blending modern technology with natural splendor.

http://m.acs.cn/3u4138/

Baan Hinta, Koh Samui

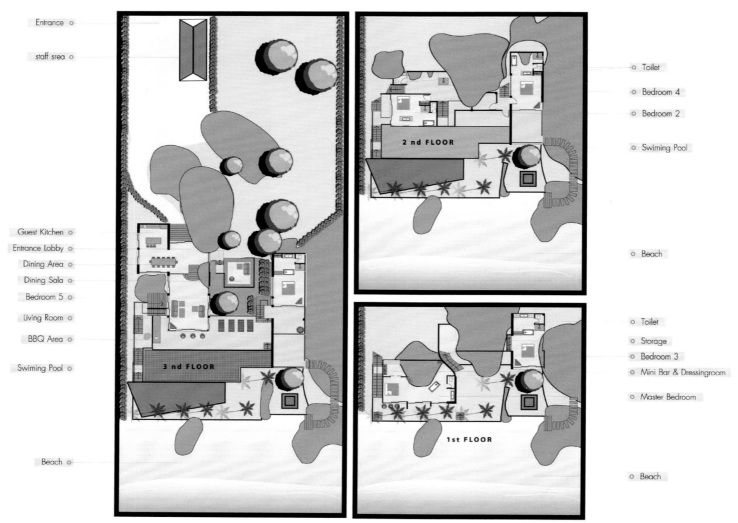

巴安欣塔别墅沿美丽的沙滩成排延伸，地理位置优越。标志性的屋顶结构使主生活区空间高达10米，十分开阔。从室内到室外，每个空间都很奢华，而且拥有广阔的海景。

该建筑因其拥有五间卧室这一惊人的结构而出名，别墅还设有大型游泳池、休息大厅和独特的户外电影区。标志性的屋顶结构主导着建筑主题：用杉木瓦覆盖，陡峭的屋顶给主生活区提供了充裕的空间，高达10米，而卧室区的屋顶变矮了，显得更加亲密。浴室和户外沐浴空间运用天然岩石，给整个质朴的室内空间增添了自然笔触，实在令人赞叹。儿童卧室中高大的玻璃墙是又一个亮点，透过玻璃墙便是泳池——模拟的水族馆。

缓缓步入别墅，映入眼帘的是极为宽敞、错层式的客厅和餐厅，空间都被木材包裹起来，拥有绝佳的热带风光。从室内到室外，奢华的建筑与广阔的海滩、海洋风景融为一体，不见任何瑕疵。从设有两张大沙发和家庭娱乐设备的休息室，到装有高清户外电影设施的美丽池边露台，再到足够容纳20人的宽敞餐厅，休闲放松空间十分充裕。开放式的西式厨房内设有专为烹饪高手打造的一系列的LG、西门子电器和大型木质餐桌。

五间装饰豪华的双人卧室中摆放着大号床铺，在空间内都可享受全面的家庭娱乐，还能欣赏令人称羡的海景。浴室中设有独立式浴缸、淋浴隔间和男女盥洗台。步入式的衣柜中摆放着超级舒服的毛绒睡衣和拖鞋。其中一间主卧室有直接通往海滩的通道。在第二层中，公共露台把卧室都连接起来，而露台上则是又一间户外浴室。

生活区可以直接通往日光甲板，甲板上摆放着柔软的两用长椅。私人大水池旁设有投影仪和屏幕，现代科技与自然景观的结合给人以独一无二的户外电影体验。

CONTRIBUTORS

Awesome Villas

Awesome Villas is a motivated team of professionals who act as an exclusive representative to villa owner's of Asia's most admired and desired multi-million dollar private estates and villas. They offer travelers and international brokers who specialize in sales and rentals, the finest luxury villas available in Phuket.

When making reservations with Awesome Villas, their staff will tailor to your booking requirements to make sure the villa you booked is tailored and personalized to your requirements. Guests will have a fully equipped luxury villa with a knowledgeable Villa Manager, housemaids, a personalized chef service to fit any dietary needs and private concierge service to help you with any information or special requests that you may have.

Banyan Tree Ungasan, Bali

Banyan Tree Ungasan features 73 spacious pool villas, set in their own lush landscaped garden with outdoor jet pool, outdoor shower, marble bathroom, and a Balinese balé (outdoor pavilion). The highlight of each modern villa is the generously-sized private infinity pool, arguably the most expansive in Bali, which starts from 10 metres in length. The property comprises a range of one, two and three-bedroom villas. Living area and bedroom are connected via courtyard, and carefully selected furnishings combine contemporary design with touches of Balinese artistry.

Designed by Peddle Thorp Architects with the interior décor by Wilson Associates Inc, Banyan Tree Ungasan was envisioned as a symbiosis of old and new, antiquity and modern. Traditional principles and local considerations were used to maximise the unparalleled ocean views while reflecting the arrangements of a Balinese village. The entry to the villa is across a terrazzo stone pavement that appears to float over a lotus pond, leading to a living room with a five-metre pitch ceiling.

Ju-Ma-Na restaurant, which means "silver pearl" in Arabic, offers modern gourmet menus. Observe the chefs at work in the display kitchen, or opt for the dramatic cliff-edge scenery, a superb setting for the excellent cuisine. The adjacent Ju-Ma-Na Bar offers sundowners along with Arabic finger food and Moroccan shishas (water pipes).

Tamarind, set in the tranquil surroundings of the Banyan Tree Spa, will serve a variety of organic dishes, alongside a menu of Jamu herbal drinks, and Bambu presents popular international and local specialties in a spacious garden setting, sometimes with live entertainment.

Firefly Collection

Firefly Collection is luxury travel specialist, showcasing a personally selected portfolio of the most exclusive luxury ski chalets and luxury villas around the world.

Firefly Collection is about quality, not quantity. Each property is individually assessed according to our strict standards of high quality luxury accommodation and exceptional service, and all are included on merit alone. Only the best luxury villas and luxury ski chalets are invited to become part of Firefly Collection.

Think of it as your shortlist of the very best properties, all in one place.

Hunter Sotheby's International Realty

Hunter Sotheby's International Realty was established in 2006 and is the exclusive representative of the prestigious Sotheby's International Realty organisation in Phuket and Southern Thailand. To those who value the unique, Hunter Sotheby's International Realty is the local real estate services provider that offers unrivaled access to qualified people and distinctive properties around the world. We specialise in luxury lifestyle properties and unique opportunities, offering unparalleled, discreet, professional services to our clients and are dedicated to the beautiful and exotic region of Southern Thailand. This area of South East Asia offers the perfect, idyllic tropical destination to live, invest or second home.

Our real estate professionals work together with you to serve your needs and requirements so that your property purchase can be undertaken seamlessly, in complete confidence and without complication. Our impartiality is paramount and sacrosanct.

Unique Retreats

Welcome to Unique Retreats, a portfolio of luxury holiday properties with outstanding location, facilities and services.

Strong with over a decade of field experience in the management and rental of high end properties, we continuously leverage our operational understanding of front-line and behind the scenes of luxury villa representation, as well as the current rates and trends within the vacation rental market.

Combining the market awareness about our villa and apartment label "Unique Retreats" while driving the attention to each individual property brand, we provide private owners and managers a global Internet strategy with which to promote their properties in a consistent and comprehensive manner.

Luxury rental holidays are increasingly renown as alternatives to 5-star and boutique hotels thanks to the opportunity of being able to enjoy full privacy, greater facilities, very personalized service and stunning location. Considering the increasing choice of holiday rentals made available, Unique Retreats ensures to carefully select properties which would only stand out of the ordinary and represent the ideal mix of what savvy travellers would seek.

Our team, in close cooperation with the property owners and managers, provides sales & marketing representation services, from creating the property brand to negotiating contracts with booking agents and handling reservations.

Zekkei Collection

The Zekkei Collection story began as their founders got behind the idea to build the ultimate venue for entertaining business contacts and hosting regular family get-togethers in the great resort community of Koh Samui. The team went on to develop over 20 luxury villas and chalets in popular vacation destinations throughout Asia.

Zekkei Collection's impressive portfolio now comprises only the finest holiday properties in vacationer-friendly cities including Phuket, Koh Samui, Niseko and Bali, all of which the team has personally inspected and would genuinely recommend basing on over 10 years' experience in high-end real estate. Zekkei has a substantial development pipeline with upcoming properties beyond Asia.

Doubling as the developer and marketer of luxury tropical villas and ski chalets, their team does the legwork for guests by screening for the best private holiday homes. Members enjoy the lowest market rental prices booking with us, as well as travel trends and destination news by following our online magazine Zekkei Lifestyle.

ARTPOWER

Acknowledgements
We would like to thank all the designers and companies who made significant contributions to the compilation of this book. Without them, this project would not have been possible. We would also like to thank many others whose names did not appear on the credits, but made specific input and support for the project from beginning to end.

Future Editions
If you would like to contribute to the next edition of Artpower, please email us your details to: artpower@artpower.com.cn